Collins

11+

Arithmetic

Practice Workbook

Faisal Nasim

Contents

ACKNOWLEDGEMENTS

The author and publisher are grateful to the copyright holders for permission to use quoted materials and images.

Every effort has been made to trace copyright holders and obtain their permission for the use of copyright material. The author and publisher will gladly receive information enabling them to rectify any error or omission in subsequent editions. All facts are correct at time of going to press.

Published by Collins
An imprint of HarperCollins*Publishers*
1 London Bridge Street
London SE1 9GF

1st Floor, Watermarque Building, Ringsend Road
Dublin 4, Ireland

ISBN: 9781844199211

First published 2018
This edition published 2020
Previously published as Letts

10 9 8 7 6 5 4

© HarperCollins*Publishers* Limited 2020

All rights reserved. No part of this publication may be reproduced, stored in a retrieval system, or transmitted, in any form or by any means, electronic, mechanical, photocopying, recording or otherwise, without the prior permission of Collins.

British Library Cataloguing in Publication Data.

A CIP record of this book is available from the British Library.

Author: Faisal Nasim
Commissioning Editor: Michelle I'Anson
Editor and Project Manager: Sonia Dawkins
Cover Design: Kevin Robbins and Sarah Duxbury
Text Design, Layout and Artwork: Ian Wrigley
Production: Natalia Rebow
Printed and bound in the UK using 100% Renewable Electricity at CPI Group (UK) Ltd

Please note that Collins is not associated with CEM in any way. This book does not contain any official questions and it is not endorsed by CEM.

Our question types are based on those set by CEM, but we cannot guarantee that your child's actual 11+ exam will contain the same question types or format as this book.

Guidance for Parents

About the 11+ tests

In most cases, the 11+ selection tests are set by GL Assessment (NFER), CEM or the individual school. You should be able to find out which tests your child will be taking on the website of the school they are applying to, or from the local authority.

The CEM test usually consists of two papers and in each paper pupils are tested on their abilities in verbal, non-verbal and numerical reasoning. Tests are separated into small, timed sections accompanied by audio instructions. It appears the content required for CEM tests changes from year to year, and the level at which pupils are tested ranges from county to county.

For pupils to do well in the CEM tests:

- they must have strong arithmetic skills
- they must have strong reasoning and problem-solving skills
- they must have a strong core vocabulary
- they must be flexible and able to understand and respond to a wide range of question types and formats, without being panicked by unfamiliar question types
- they must be able to work under time pressure.

About this book and how to use it

This book provides preparation for the numerical reasoning part of the exam and, more specifically, arithmetic questions. These tests will help students to develop the arithmetic fundamentals that will allow them to work quickly and accurately in the actual exam.

Helpful tips are given for questions where students commonly make mistakes – encourage your child to follow these tips to help ensure they don't lose vital marks in the exam.

The book contains:

- 30 tests focused on building arithmetic skills in preparation for the CEM 11+ test
- Helpful tips to avoid losing marks
- A complete set of answers
- Score sheet to track results over multiple attempts.

We suggest that your child writes the answers on a separate piece of paper when attempting the tests for the first time, as you may want your child to complete them again at a later date.

You should record your child's first attempt at each test in the table at the end of the book. At a later date, your child should repeat any section where the score was poor or where they were unable to finish.

The time allowed for each test is set to be challenging, but eventually manageable for your child.

Depending on the amount of practice of timed tests which your child has completed prior to using this book, initially your child may find the tests difficult to complete. However, it is through practice of timed tests that pupils gain more confidence and become more time-aware.

Your child will need:

- A quiet place to do the tests
- Paper (on which to write the answers)
- A clock/watch which is visible to your child
- A pencil.

Your child should **not** use a calculator for any of these tests.

Practice will help your child to do his or her best on the day of the tests. Working through the more difficult question types allows your child to practise answering a range of test-style questions that will help them achieve the best results. It also provides an opportunity for them to learn how to manage time effectively, so that time is not wasted during the test and any 'extra' time is used constructively for checking.

Test 1

 You have 6 minutes to complete the following section.
You have 12 questions to complete within the time given.

Write the correct answer in the boxes provided (one digit per box).

EXAMPLE

Calculate (7 × 9) + (6 × 8)

The answer is:

$\boxed{1}\ \boxed{1}\ \boxed{1}$

(7 × 9) + (6 × 8) = 63 + 48 = 111

---

(1) Calculate 4.5 km ÷ 2

$\boxed{}\ .\ \boxed{}\ \boxed{}$ km

(2) Calculate 0.83 × 7

$\boxed{}\ .\ \boxed{}\ \boxed{}$

(3) Calculate $\frac{1}{4} \times \frac{1}{3}$

$\dfrac{\boxed{}}{\boxed{}\ \boxed{}}$

(4) x × 25 g = 1.5 kg

What is the value of x?

$\boxed{}\ \boxed{}$

HELPFUL TIP: Make sure all the numbers in a sum are denominated in the same unit before calculating your answer.

5. Seventy thousand four hundred and nine.

   Write this number in digits.

   ⬜⬜,⬜⬜⬜

6. How many square numbers are greater than 18 but less than 75?

   ⬜

7. How many minutes are there in $4\frac{5}{6}$ hours?

   ⬜⬜⬜ minutes

8. 7 apples cost a total of £1.47.

   What is the cost of 4 apples?

   ⬜⬜ p

9. David pays £29 per month for his mobile phone.

   How much does he pay annually?

   £⬜⬜⬜

10. A dress costs £32.

    The price of the dress increases by 20%.

    What is the new price of the dress?

    £⬜⬜.⬜⬜

11. A square has a perimeter of 44 cm.

    What is the width of the square?

    ⬜⬜⬜ mm

12. A circle is divided into 9 equal segments.

    What is the size of the angle between
    the 2 straight sides in each segment?

    ⬜⬜ °

Test 2

 You have 6 minutes to complete the following section.
You have 12 questions to complete within the time given.

Write the correct answer in the boxes provided (one digit per box).

EXAMPLE

1 orange costs £0.65. What is the cost of 6 oranges?

The answer is:

| 3 | 9 | 0 | p |

£0.65 = 65 p
65 p × 6 = 390 p

---

(1) Calculate 56 × 34

☐ , ☐ ☐ ☐

(2) What is the square root of 100?

☐ ☐

(3) Calculate $\frac{2}{3} - \frac{1}{4}$

Write your answer as a fraction in its lowest terms.

$\frac{\square}{\square}$

HELPFUL TIP: When adding and subtracting fractions, you first have to find a common denominator.

(4) Calculate 44 m ÷ 11

☐ , ☐ ☐ ☐ mm

Page 6

(5) How many factors does the number 36 have?

(6) Which number comes next in this sequence?

25, 36, 49, 64, ?

(7) What is the perimeter of a square with an area of 4 cm²? ☐ cm

(8) What is 96% of £25? £ ☐☐

(9) The radius of Circle A is three times the diameter of Circle B.

Four times the radius of Circle B is 50 cm.

What is the diameter of Circle A? ☐.☐ m

(10) Bob runs 400 m in 50 seconds.

What is Bob's speed in kph? ☐☐.☐ kph

(11) 4 chicks are born every 20 minutes.

How many chicks are born between 00:25 and 02:45? ☐☐

(12) Point Y has the coordinates (1, 6).

It is translated 3 squares up and 5 squares left.

What are the new coordinates of Point Y? (-☐,☐)

Test 3

 You have **6 minutes** to complete the following section.
You have **12 questions** to complete within the time given.

Write the correct answer in the boxes provided (one digit per box).

EXAMPLE

One quarter of the students in Class A have brown eyes. If there are
52 students in Class A, how many do not have brown eyes?

The answer is:

| 3 | 9 |

Number of students who have brown eyes = 52 ÷ 4 = 13
Number of students who do not have brown eyes = 52 − 13 = 39

---

① Calculate 342 × 82

② Calculate 120 ÷ 8

③ $V = 7$ $F = 9$

What is the value of $3V + 5F$?

④ Round 34.2829 to 2 decimal places.

⑤ Add the smallest 2-digit number to the smallest 3-digit number.

6 What is the perimeter of a regular hexagon with a side length of 2.35 cm?

☐☐.☐ cm

7 2 angles in a triangle are 67° and 24°.

What is the size of the 3rd angle in the triangle?

☐☐ °

8 A bowl of sweets is shared between Martin and Ben in the ratio 8:7.

Ben receives 56 sweets.

How many sweets were there in the bowl in total before they were shared?

☐☐☐

9 What is the order of rotational symmetry of a rectangle?

☐

10 How many numbers are factors of both 36 and 48 but not factors of 54?

☐

11 A child's theatre ticket costs $\frac{2}{7}$ as much as an adult ticket.

An adult ticket costs £2.17.

What is the total cost of 2 child tickets and 3 adult tickets?

£☐.☐☐

12 John and his brothers each have 4 children. $\frac{1}{4}$ of these children are male.

If John has 4 brothers, how many of the children are female?

☐☐

Test 4

Write the correct answer in the boxes provided (one digit per box).

EXAMPLE

Calculate $(7 \times 9) + (6 \times 8)$

The answer is:

1 1 1

$(7 \times 9) + (6 \times 8) = 63 + 48 = 111$

---

1. Write $23\frac{2}{3}$ as a top-heavy, improper fraction.

2. Calculate $4{,}928 + 3{,}293$

3. Calculate $71 - (21 \times 3)$

4. Calculate $5 \div \frac{1}{4}$

5. Decrease $45{,}380$ by 10%.

6) A festival lasts from 17th March to 2nd April, inclusive.

For how many days does the festival last?

☐☐ days

7) What is the height of a triangle with a base measuring 12 cm and an area of 36 cm²?

☐ cm

8) 7 children each drink 330 ml of cola.

How much do they drink in total?

☐.☐☐ l

9) Pia must take 10 ml of medicine in the morning and 15 ml of medicine in the evening. The medicine is packaged in capsules, each containing 5 ml.

How many capsules does Pia consume in a fortnight?

☐☐

10) Teabags cost 14 p each or can be bought as a pack of 9 for £1.10.

Frieda buys exactly 100 teabags.

What is the smallest possible amount she could have paid?

£☐☐.☐☐

11) Arun bought a £22 game with his savings. This resulted in an 11% drop in his total savings.

How much money did Arun have in savings before he bought the game?

£☐☐☐

12) The temperature on Tuesday was 10° less than on Monday. The temperature on Thursday was 3° more than on Tuesday.

If the temperature on Monday was 29°, what was the temperature on Thursday?

☐☐ °

Test 5

 You have 6 minutes to complete the following section.
You have 12 questions to complete within the time given.

Write the correct answer in the boxes provided (one digit per box).

EXAMPLE

1 orange costs £0.65. What is the cost of 6 oranges?

The answer is:

$\boxed{3}\boxed{9}\boxed{0}$ p

£0.65 = 65 p
65 p × 6 = 390 p

---

(1) Round 322,283 to the nearest 10 thousand.

$\boxed{}\boxed{}\boxed{},\boxed{}\boxed{}\boxed{}$

(2) Calculate 195 ÷ 15

$\boxed{}\boxed{}$

(3) Calculate $\frac{7}{10} \times \frac{4}{8}$

Write your answer as a fraction in its lowest terms.

$\dfrac{\boxed{}}{\boxed{}\boxed{}}$

(4) Write $5\frac{3}{5}$ as a decimal.

$\boxed{}.\boxed{}$

(5) A petrol tank currently holds $12\frac{1}{5}$ litres of petrol and is $\frac{1}{3}$ full.

How much petrol can the tank hold in total?

$\boxed{}\boxed{}.\boxed{}$ l

(6) Multiply the largest 2-digit square number by 7.

[ ][ ][ ]

(7) What is the radius of a circle with a diameter of 17.856 cm?

[ ][ ].[ ][ ] mm

(8) A car travels 45 km every half an hour.

How far will the car travel in $4\frac{1}{4}$ hours?

[ ][ ][ ].[ ] km

HELPFUL TIP: *Think about how speed, distance and time are related.*

(9) $\frac{1}{2}$ the toads in a pond are green and $\frac{1}{4}$ of these green toads have yellow spots.

If there are 128 toads in the pond in total, how many of them are green but without yellow spots?

[ ][ ]

(10) A map is drawn to a scale of 1:1,000,000.

What is the actual length of a distance of 3 cm on the map?

[ ][ ] km

(11) What is the highest common factor of 96, 84 and 60?

[ ][ ]

(12) The cost of using the internet in a hotel is calculated as £3 per day plus £0.01 for every minute used.

Ian stayed at the hotel for 1 week and used the internet for half an hour per day.

How much did he pay to use the internet in total?

£[ ][ ].[ ][ ]

Test 6

INSTRUCTIONS

 You have 6 minutes to complete the following section.
You have 12 questions to complete within the time given.

Write the correct answer in the boxes provided (one digit per box).

EXAMPLE

One quarter of the students in Class A have brown eyes. If there are
52 students in Class A, how many do not have brown eyes?

The answer is:

| 3 | 9 |

Number of students who have brown eyes = 52 ÷ 4 = 13
Number of students who do not have brown eyes = 52 − 13 = 39

---

(1) $T - 45.2 = 68.9$

What is the value of T?

(2) Write 62.5% as a fraction in its lowest terms.

(3) Round 10.0077 to the nearest hundredth.

(4) Calculate 4,829 × 73

(5) The price of a hat is reduced in a sale by 20%.

The sale price of the hat is £96.

What was the original price of the hat?

£

(6) 1 angle in a right-angled triangle is 27°.

What is the size of the smallest angle in the triangle?

◻◻ °

(7) How many square tiles with a side length of 50 cm are needed to cover a square floor with a side length of 3 metres?

◻◻

(8) A rocket travels 2.5 km in 15 seconds.

What is the rocket's speed in kph?

◻◻◻ kph

(9) Pete has an equal number of five-pence and two-pence coins in his pocket that have a total value of £1.33.

How many five-pence coins does Pete have in his pocket?

◻◻

(10) On Tuesday, the price of a plane ticket drops by 10%. On Wednesday, the price of the plane ticket increases by 10% compared to the reduced price.

The original price of the plane ticket, before it was reduced, was £350.

What was the price of the plane ticket after the 10% increase on Wednesday?

£ ◻◻◻ . ◻◻

(11) A square with an area of 81 cm² is cut in half to create 2 identical rectangles.

What is the total perimeter of the 2 rectangles?

◻◻ cm

(12) The size of Angle A in Triangle B is 57°.

Triangle B is enlarged by a scale factor of 2.

What is the size of Angle A in the enlarged triangle?

◻◻ °

Test 7

 You have 6 minutes to complete the following section.
You have 12 questions to complete within the time given.

Write the correct answer in the boxes provided (one digit per box).

EXAMPLE

Calculate $(7 \times 9) + (6 \times 8)$

The answer is:

| 1 | 1 | 1 |

$(7 \times 9) + (6 \times 8) = 63 + 48 = 111$

---

1. Calculate $9^2 + 4^2$

2. Calculate 832×93

3. Fourteen million and six.

   Write this number in digits.

4. Calculate $56 \text{ kg} \div 100$ [ ][ ][ ] g

HELPFUL TIP: *Before making any calculations, check if your answer needs to be denominated in a specific unit.*

(5) What is the probability of rolling an even number with a fair dice?

Write your answer as a fraction in its lowest terms.

$\dfrac{\boxed{}}{\boxed{}}$

(6) 2 kg of cement costs £15.20.

What is the cost of 600 g of cement?

£ $\boxed{}$. $\boxed{}\boxed{}$

(7) What is the perimeter of an equilateral triangle with a side length of 22.8 cm?

$\boxed{}\boxed{}$. $\boxed{}$ cm

(8) A cargo container weighs 455 kg.

What is the total weight of 7 cargo containers?

$\boxed{}$. $\boxed{}\boxed{}\boxed{}$ tonnes

(9) The ratio of one-pence, two-pence, five-pence and ten-pence coins in a jar is $3:4:2:5$. There are 98 coins in the jar in total.

How many of the coins in the jar have an even value?

$\boxed{}\boxed{}$

(10) Tim buys 4 packets of sweets, each costing £1.21.

He pays with a note and receives £15.16 in change.

What was the value of the note that Tim used to pay for his sweets? £ $\boxed{}\boxed{}$

(11) Richard spends $\frac{1}{3}$ of his week asleep.

How many hours does Richard sleep per week?

$\boxed{}\boxed{}$ hours

(12) 5 spiders can spin 3 webs in 6 minutes.

How many webs can 15 spiders spin in 1 hour?

$\boxed{}\boxed{}$

Test 8

INSTRUCTIONS

 You have 6 minutes to complete the following section.
You have 12 questions to complete within the time given.

Write the correct answer in the boxes provided.

EXAMPLE

1 orange costs £0.65. What is the cost of 6 oranges?

The answer is:

| 3 | 9 | 0 | p |

£0.65 = 65 p
65 p × 6 = 390 p

---

(1) Calculate 122 cm × 8

☐.☐☐ m

(2) $\frac{10}{W} = \frac{110}{77}$

What is the value of W?

☐

(3) Calculate £7.44 ÷ 8

☐☐ p

(4) 1.43 million

Write this number in digits.

☐,☐☐☐,☐☐☐

(5) Sarah faces north and turns 135° clockwise.

In which direction does Sarah now face?

☐

(6) Rob leaves home and returns $1\frac{1}{3}$ hours later at 5:24 p.m.

At what time did Rob leave home?

☐ : ☐☐ p.m.

(7) What is the largest 2-digit cube number?

☐☐

(8) A rectangle has a length of 7 cm and a width that is half of its length.

What is the area of the rectangle?

☐☐.☐ cm²

(9) A square and a circle have the same area.

If the square has a perimeter of 18 cm, what is half the area of the circle?

☐☐.☐☐☐ cm²

(10) Uma thinks of a shape with 2 pairs of equal sides, 1 line of symmetry and no pairs of parallel sides.

What shape is Uma thinking of?

☐

(11) If the volume of a cube is 1,000 cm³, what is the length of 1 side of the cube?

☐☐ cm

HELPFUL TIP: *Think about how the side length of a cube relates to its volume.*

(12) What is $\frac{2}{3}$ of $\frac{1}{2}$ of 0.8 of 75% of 1,000?

☐☐☐

Test 9

 You have 6 minutes to complete the following section.
You have 12 questions to complete within the time given.

Write the correct answer in the boxes provided.

EXAMPLE

One quarter of the students in Class A have brown eyes. If there are
52 students in Class A, how many do not have brown eyes?

The answer is:

3	9

Number of students who have brown eyes = 52 ÷ 4 = 13
Number of students who do not have brown eyes = 52 − 13 = 39

---

(1) Calculate 973 seconds − 613 seconds ☐ minutes

(2) Calculate 4,928 × 88 ☐☐☐,☐☐☐

(3) What is twice the square root of 49? ☐☐

(4) 30% of C = 60

What is the value of C? ☐☐☐

(5) What is the probability of rolling a prime number with a fair dice?

Write your answer as a fraction in its lowest terms.

☐
☐

(6) If 16ᵗʰ October is a Friday, what day is 7ᵗʰ November in the same year?

(7) 7 litres are drained from a water tank every 15 minutes.

If there are 224 litres in the tank, how long will it take for all the water to be drained?

☐ hours

(8) The coordinates of 3 corners of a rectangle are (4, 8), (2, 8) and (4, 13).

What are the coordinates of the other corner of the rectangle?

(☐ , ☐☐)

(9) A cuboid has a length of 10 cm, a width of 3 cm and a height of 2 cm.

Oliver paints half of each face of the cuboid red.

What surface area of the cuboid is not painted red?

☐☐ cm²

(10) What is the order of rotational symmetry of this figure?

IOIOI

☐

(11) 1 cm on a map represents an actual distance of 5.5 km.

What actual distance is represented by 1 mm on the map?

☐☐☐ m

(12) 15 black cards and 12 red cards are placed in a sack.

1 black card is removed and then a card is selected at random.

What is the probability that the selected card is red?

Write your answer as a fraction in its lowest terms.

$\frac{\square}{\square\square}$

Test 10

 You have 6 minutes to complete the following section.
You have 12 questions to complete within the time given.

Write the correct answer in the boxes provided (one digit per box).

EXAMPLE

Calculate $(7 \times 9) + (6 \times 8)$

The answer is:

| 1 | 1 | 1 |

$(7 \times 9) + (6 \times 8) = 63 + 48 = 111$

① Calculate £8.72 × 18

£ ☐ ☐ ☐ . ☐ ☐

② Calculate 7 km − 374 m

☐ , ☐ ☐ ☐ m

③ How many eighths are there in $5\frac{5}{8}$?

☐ ☐

④ Calculate 7 weeks − 13 days

☐ ☐ days

⑤ Write a number that is greater than 3.45 but less than 3.46.

☐ . ☐ ☐ ☐

(6) Group F consists of all the even whole numbers greater than 18 but less than 31.

How many numbers are in Group F?

(7) 4 out of every 9 animals in a park are dogs.

80 animals in the park are not dogs.

How many animals are there in the park in total?

(8) What is the area of a rectangular table with a length of 13 m and a width of 0.8 m?

$\boxed{}\boxed{}.\boxed{}$ m²

(9) Ewan is 8th from the back of a queue and 12th from the front.

How many people are in the queue?

(10) Otto's finishing time in a race was 43 seconds.

Toby finished 4 seconds faster than Otto.

Alan finished 8 seconds slower than Toby.

What was the average finishing time of Otto, Toby and Alan?

$\boxed{}\boxed{}$ seconds

HELPFUL TIP: *Remember that if someone finishes a race faster, it will take them less time, not more!*

(11) $*$ is a function. $y* = 4y + \frac{y}{2}$

What is the value of $6*$?

(12) Which of these numbers is closest in value to 0.05?

0.051, 0.005, 0.055, 0.493, 0.048, 0.045

$\boxed{}.\boxed{}\boxed{}\boxed{}$

Test 11

INSTRUCTIONS

 You have 6 minutes to complete the following section.
You have 12 questions to complete within the time given.

Write the correct answer in the boxes provided (one digit per box).

EXAMPLE

1 orange costs £0.65. What is the cost of 6 oranges?

The answer is:

| 3 | 9 | 0 | p |

£0.65 = 65 p
65 p × 6 = 390 p

---

(1) $9Q - 647 = 433$

What is the value of Q?

☐☐☐

(2) Calculate $564 \div 12$

☐☐

(3) Calculate $7.382 - 2.484$

☐.☐☐☐

(4) Calculate $11^2 + (4^2 - 3^2)$

☐☐☐

HELPFUL TIP: Remember to solve the sum in the brackets first.

(5) Paul writes down all the even prime numbers.

How many numbers does Paul write down?

☐

(6) A model of a train is built on a 1:8 scale.

If the actual train is 10.4 metres long, how long is the model? ☐☐☐ cm

(7) What number is 17.453 less than double 34.29? ☐☐.☐☐☐

(8) 2 angles in a quadrilateral are 93° and 120°.

What is the sum of the other angles in the quadrilateral? ☐☐☐ °

(9) The population of Rinnes is 789,000, rounded to the nearest 1,000.

What is the largest possible population of Rinnes? ☐☐☐,☐☐☐

(10) A regular octagon has the same perimeter as a regular hexagon.

If the hexagon has a side length of 8 cm, what is the total length of 2 sides of the octagon? ☐☐ cm

(11) 85% of the participants in a survey stated that they enjoyed running.

If 105 people who took the survey did not state that they enjoyed running, how many people took the survey in total? ☐☐☐

(12) Wendy must replace 50% of the seats in her fleet of 11 vans.

If each van has 12 seats and the cost to replace 2 seats is £35.50, how much must Wendy pay in total? £☐,☐☐☐.☐☐

Test 12

 INSTRUCTIONS

 You have 6 minutes to complete the following section.
You have 12 questions to complete within the time given.

Write the correct answer in the boxes provided (one digit per box).

EXAMPLE

One quarter of the students in Class A have brown eyes. If there are
52 students in Class A, how many do not have brown eyes?

The answer is:

| 3 | 9 |

Number of students who have brown eyes = 52 ÷ 4 = 13
Number of students who do not have brown eyes = 52 − 13 = 39

---

(1) How many millimetres are there in 2.6 m?

☐,☐☐☐ mm

(2) Calculate 1.4 l + 345 ml + 211 ml

☐.☐☐☐ l

(3) Calculate 333 × 33

☐☐,☐☐☐

(4) Calculate $\frac{1}{2} + \frac{3}{5}$

Write your answer as a decimal.

☐.☐

(5) The temperature increases from −18 °C by 22 °C.

What is the temperature now?

☐ °C

6) Subtract the number of months in a year with 31 days from the number of leap years from 1921 to 1969.

☐

HELPFUL TIP: *Remember that leap years are always multiples of 4.*

7) Rachel cuts a 3.5 m piece of ribbon into 50 equal pieces.

What is the total length of 3 pieces?

☐☐☐ mm

8) Yasmin thinks of a number, divides it by 2, adds 7 and then multiplies by 3.

The answer is 48.

What number did Yasmin think of?

☐☐

9) £1 has the same value as 1.15 euros. Henry buys a book for 9.20 euros.

What is the value of the book in pounds?

£ ☐

10) How much greater is −1,293 than −1,357?

☐☐

11) Last year, Belle was 4 times as old as Fred.

If Fred will be 7 next year, how old will Belle be next year?

☐☐

12) How many faces does a pentagonal prism have?

☐

Test 13

 You have 6 minutes to complete the following section.
You have 12 questions to complete within the time given.

Write the correct answer in the boxes provided (one digit per box).

EXAMPLE

Calculate $(7 \times 9) + (6 \times 8)$

The answer is:

| 1 | 1 | 1 |

$(7 \times 9) + (6 \times 8) = 63 + 48 = 111$

---

① Calculate 1.5 km × 5

◻ , ◻ ◻ ◻ m

② Calculate £89.32 − £87.22

◻ ◻ ◻ p

③ $\frac{T}{33} = \frac{0}{12}$

What is the value of T?

◻

④ Calculate 65,283 + 40,002

◻ ◻ ◻ , ◻ ◻ ◻

⑤ 1 lion eats 2 kg of meat in half an hour.

How long will it takes 3 lions to eat 18 kg of meat?

◻ ◻ minutes

(6) Which of these numbers is the largest?

−45,433, −45,434, −45,431, −45,436

−⬜⬜,⬜⬜⬜

HELPFUL TIP: *The largest negative number will be the one that is closest to zero.*

(7) Neil paints $\frac{2}{3}$ of the faces of a cube with blue paint.

How many faces of the cube are not painted blue?

⬜

(8) What is the volume of a swimming pool with a length of 25 m, a width of 10 m and a depth of 2 m?

⬜⬜⬜ m³

(9) $(7C + 8) \times C = 18 \times (D − D)$

D is a positive number.

What is the value of C?

⬜

(10) A map is drawn so that 1 mm on the map represents an actual distance of 25 km.

What distance on the map represents an actual distance of 1,000 km?

⬜ cm

(11) A rectangle with a length of 8 cm and a width of 2 cm is cut along its lines of symmetry to create four smaller identical rectangles.

What is the total perimeter of these four smaller rectangles?

⬜⬜⬜ mm

(12) A submarine is currently 350 metres below sea level. It fires a flare in a straight line to reach the height of a helicopter that is 757 metres directly above it.

How many kilometres above sea level is the helicopter?

⬜.⬜⬜⬜ km

Test 14

 You have 6 minutes to complete the following section.
You have 12 questions to complete within the time given.

Write the correct answer in the boxes provided.

EXAMPLE

1 orange costs £0.65. What is the cost of 6 oranges?

The answer is:

[3][9][0] p

£0.65 = 65 p
65 p × 6 = 390 p

---

(1) Calculate 0.83 × 12

☐.☐☐

(2) What is half of $\frac{1}{8}$ of 64?

☐

(3) 65% × 200 = $\frac{R}{2}$

What is the value of R?

☐☐☐

(4) Calculate 432 × 83

☐☐,☐☐☐

(5) 2 apples and 3 lemons weigh a total of 0.8 kg.

If an apple weighs 220 g, what is the total weight of 2 lemons?

☐.☐☐ kg

(6) What is the largest number of coins that can be used to make a total of £4.56?

(7) The time shown by a clock is 5:15 p.m.

What is the time after the minute hand turns through 810°?

◯ : ◯◯ p.m.

(8) The ratio of blue buttons to red buttons in a pot is 4:7.

If there are 16 blue buttons in the pot, how many more red buttons are there in the pot than blue buttons?

(9) James faces north. He walks for 2 km and then turns 90° anticlockwise

He then walks for 1 km and turns 225° anticlockwise.

In which direction does James now face?

(10) 50% of a rectangular board is painted green, $\frac{1}{5}$ is painted red and the rest is painted blue. A dart is thrown on the board at random.

What is the probability that the dart lands on a blue section of the board?

(11) A cup can hold 175 ml of water. Kamal fills as many cups as possible from a jug containing 2 litres of water.

How much water is left in the jug?

◯ . ◯◯◯ l

(12) What is the difference in seconds between $\frac{3}{5}$ of an hour and $\frac{5}{8}$ of 2 hours?

◯ , ◯◯◯ seconds

Test 15

 You have 6 minutes to complete the following section.
You have 12 questions to complete within the time given.

Write the correct answer in the boxes provided (one digit per box).

EXAMPLE

One quarter of the students in Class A have brown eyes. If there are
52 students in Class A, how many do not have brown eyes?

The answer is:

| 3 | 9 |

Number of students who have brown eyes = 52 ÷ 4 = 13
Number of students who do not have brown eyes = 52 − 13 = 39

---

(1) Calculate 15% of 2 m   ☐☐ cm

(2) How many days are there in 4 fortnights?   ☐☐ days

(3) Calculate 45 hours ÷ 6   ☐ h ☐☐ min

(4) Calculate $\frac{5}{6} - \frac{1}{6} + \frac{1}{8}$ $\frac{\square\square}{\square\square}$

(5) How many prime numbers are greater than
18 but less than 29?   ☐

(6) How much change do I receive from £10 if I buy 3 pens, each costing £0.93?

☐☐☐ p

(7) Henry eats $\frac{1}{2}$ of a cake and Finn eats $\frac{1}{2}$ of the remainder.

What fraction of the cake remains?

$\frac{\boxed{}}{\boxed{}}$

(8) A side of a rectangle is enlarged by a scale factor of 8.

The enlarged side is 5.36 metres long.

How long was the side before it was enlarged?

☐☐☐ mm

(9) What is the median of the following numbers?

65, 45, 32, 87, 12, 43, 23, 76

☐☐

HELPFUL TIP: *To help you calculate the median, first place the numbers in size order.*

(10) The cost of a pen is 27 p. Henrietta buys some pens and pays with a £10 note. She receives £5.14 in change.

How many pens does she buy?

☐☐

(11) Train A leaves Station C at 14:15 and travels for 2 hours at a speed of 43 kph.

Train B leaves Station C at 13:16 and travels at a speed of 64 kph until 15:46.

How much further does Train B travel than Train A?

☐☐ km

(12) Triangle A has a base of 15 cm and an area of 225 cm².

Triangle B has double the base and three times the height of Triangle A.

What is the area of Triangle B?

☐,☐☐☐ cm²

Test 16

INSTRUCTIONS

You have 6 minutes to complete the following section.
You have 12 questions to complete within the time given.

Write the correct answer in the boxes provided (one digit per box).

EXAMPLE

Calculate (7 × 9) + (6 × 8)

The answer is:

[1] [1] [1]

(7 × 9) + (6 × 8) = 63 + 48 = 111

---

(1) Calculate 753 × 8.2

☐,☐☐☐.☐☐

(2) Calculate 74 mm × 8

☐☐.☐ cm

(3) Calculate 14 km − (234 m + 1.2 km)

☐☐.☐☐☐ km

(4) Calculate 543 ÷ 15

☐☐.☐

(5) Which number comes next
in this sequence?

786.32, 78.632, 7.8632, 0.78632, ?

☐.☐☐☐☐☐

6 A regular hexagon is formed from 6 equilateral triangles.

The perimeter of each triangle is 27.3 cm.

What is the perimeter of the hexagon?

☐☐.☐ cm

7 What is the difference between 7 mm and 12.2 m?

☐,☐☐☐.☐ cm

8 The price of a share increases by 10% every year.

If the current price of the share is £1, what will the price be in 2 years' time?

£☐.☐☐

9 A £10 note has a thickness of 0.01 mm.

What is the total thickness of a pile of £10 notes with a total value of £870?

Round your answer to 2 decimal places.

☐.☐☐ cm

10 A rectangular playing field is represented on a map with a length of 5 mm and a width of 2 mm.

The scale of the map is 1 cm to 100 m.

What is the area of the actual playing field?

☐,☐☐☐ m²

HELPFUL TIP: *Make sure both numbers in a scale ratio are denominated in the same unit before using it to make a calculation.*

11 The point (−3, −8) is reflected in the *y*-axis.

What are the coordinates of the reflected point?

(☐,−☐)

12 3 builders can construct a wall in 30 minutes.

How long would it take 1 builder to construct the wall?

☐☐ minutes

Test 17

INSTRUCTIONS

 You have 6 minutes to complete the following section.
You have 12 questions to complete within the time given.

Write the correct answer in the boxes provided (one digit per box).

EXAMPLE

1 orange costs £0.65. What is the cost of 6 oranges?

The answer is:

| 3 | 9 | 0 | p |

$£0.65 = 65$ p
65 p $\times 6 = 390$ p

---

(1) $A = 45$ $B = 2A$

What is the value of $4B$?

(2) Calculate $(\frac{1}{2}$ of $50) - (\frac{1}{5}$ of $25)$

(3) Calculate 543×28

(4) Write $\frac{45}{7}$ as a mixed fraction.

(5) A bicycle tyre has a circumference of 200 cm.

How many times will the wheel turn completely if the bike is ridden for 4 km?

(6) Half of a number is 7 less than double 18.

What is the number?

(7) The number of flies in a room doubles every 7 minutes.

If there are 7 flies in the room at 14:23, how many flies will be in the room at 14:51?

(8) How many hundredths are there in 1.34?

(9) 9 Tims have the same value as 4 Toms.

5 Toms have the same value as 18 Tums.

1 Tum has a value of 20 p.

What is the value of 3 Tims?

£ ☐ . ☐ ☐

(10) Kate wishes to bake a cake for 18 people.

Her recipe instructs her to use 4 eggs per 6 people.

Eggs are only available in boxes of 10 for £3.75.

How much must Kate spend on eggs to ensure she has enough to bake a cake for 18 people?

£ ☐ . ☐ ☐

HELPFUL TIP: *For questions involving recipes, remember that the amount of an ingredient required increases or decreases proportionately depending on the number of people served.*

(11) Which number comes next in this sequence?

−45, −34, −40, −29, −35, ?

− ☐ ☐

(12) How many times greater is $1\frac{6}{8}$ than $\frac{1}{8}$?

Test 18

INSTRUCTIONS

 You have 6 minutes to complete the following section.
You have 12 questions to complete within the time given.

Write the correct answer in the boxes provided (one digit per box).

EXAMPLE

One quarter of the students in Class A have brown eyes. If there are
52 students in Class A, how many do not have brown eyes? The answer is:

3	9

Number of students who have brown eyes = 52 ÷ 4 = 13
Number of students who do not have brown eyes = 52 − 13 = 39

---

(1) Round 45,653,348 to the nearest hundred thousand.

☐☐,☐☐☐,☐☐

(2) How many quarters are there in $4\frac{1}{2}$?

☐☐

(3) Calculate (£4.50 ÷ 5) × 6

☐☐☐ P

(4) Calculate $8^2 + 44 - 2^2$

☐☐☐

(5) How many 2-digit square numbers are also prime?

☐

(6) Zak receives £2.25 per week in pocket money.

How much does he receive annually?

£ ☐☐☐

(7) 1 angle in an isosceles triangle measures 93°.

What is the difference in size between the other
2 angles in the triangle?

☐

HELPFUL TIP: *Think about what it means for a triangle to be isosceles and how that can help you calculate angles.*

(8) A prize of £125 is divided in the ratio of 1:3:1 between
Gina, Grace and Gail. How much more did Grace receive
than Gina and Gail combined?

£ ☐☐

(9) Wendy wishes to paint a rectangular wall measuring 9 m by 6 m.

1 tin of paint costs £4.25 and can cover 10 m² with 1 coat of paint.

How much must Wendy spend on paint to cover
her whole wall with 2 coats of paint?

£ ☐☐.☐☐

(10) Point C is a translation 6 squares north and 3 squares west
of Point D. Point E is a reflection of Point C in the *x*-axis.

If Point D has the coordinates (−1, 8), what are the
coordinates of Point E?

(−☐ , −☐☐)

(11) 4 out of every 17 apples in a crate are rotten.

If there are a total of 84 rotten apples in the crate,
how many more non-rotten apples are there than rotten
apples in the crate?

☐☐☐

(12) Two equilateral triangles are placed together to
form a four-sided shape.

What is the sum of the internal angles of this four-sided shape?

☐☐☐ °

Test 19

INSTRUCTIONS

 You have 6 minutes to complete the following section.
You have 12 questions to complete within the time given.

Write the correct answer in the boxes provided (one digit per box).

EXAMPLE

Calculate $(7 \times 9) + (6 \times 8)$

The answer is:

| 1 | 1 | 1 |

$(7 \times 9) + (6 \times 8) = 63 + 48 = 111$

---

(1) How many factors does the number 80 have?

(2) $7X + 41 = 1,000$

What is the value of X?

(3) What is one thousand and eighteen less than twenty three thousand?

Write your answer in digits.

(4) How many more halves are there in $18\frac{1}{2}$ than quarters in $4\frac{3}{4}$?

(5) How much less is the sum of $\frac{1}{2}$ of 6 and $\frac{2}{3}$ of 21 than the

sum of $\frac{1}{4}$ of 1 and $\frac{7}{8}$ of 80?

6 A litre of petrol allows a motorbike to travel for 33.5 km.

How far can the motorbike travel with 4.5 litres of petrol?

☐☐☐.☐☐ km

7 The perimeter of a rectangle is 36 cm and its length is twice its width.

What is the area of the rectangle?

☐☐ cm²

8 A million buttons cost £5,324.

What is the cost, to the nearest pence, of 1 button?

☐ p

9 A lorry has 8 wheels, a car has 4 wheels and a motorbike has 2 wheels.

A car park has 5 floors. On each floor there are 7 lorries, 7 cars and 7 motorbikes.

What is the total number of wheels on all the vehicles in the car park?

☐☐☐

10 A special cell splits itself into 2 identical copies every 7 minutes.

If there are 3 cells at 15:12, how many will there be at 16:01?

☐☐☐

11 Pippa pays for a television in 11 equal instalments of £15.

The cost of buying the television in 11 instalments is 10% greater than the cost of buying it in 6 instalments.

What is the total cost of the television if it is bought in 6 instalments?

£ ☐☐☐

12 In an election, the votes for Candidates A and B were split in the ratio 52:48.

If 1,200,000 votes were cast in total, how many more votes did Candidate A receive than Candidate B?

☐☐,☐☐☐

Test 20

 You have 6 minutes to complete the following section.
You have 12 questions to complete within the time given.

Write the correct answer in the boxes provided (one digit per box).

EXAMPLE

1 orange costs £0.65. What is the cost of 6 oranges?

The answer is:

| 3 | 9 | 0 | p |

£0.65 = 65 p
65 p × 6 = 390 p

---

(1) Calculate 0.1 × 763.3

☐☐.☐☐

(2) Calculate 12,543 ml − $\frac{1}{4}$ l

☐☐.☐☐☐ l

(3) Calculate (542 × 12) − 0.1

☐,☐☐☐.☐

(4) How many even whole numbers are greater than 8 but less than 24?

☐

(5) Bob has 5 different coins in his pocket.

What is the largest possible amount of money Bob could have in his pocket?

£☐.☐☐

6) Rachel scored 45, 65, 19 and 19 in 4 tests.

What was Rachel's mean score?

7) $\frac{1}{3}$ of a bag of cherries costs 40 p.

What is the cost of $2\frac{1}{2}$ bags of cherries?

£

8) What is the time $2\frac{4}{5}$ hours before 11:59 p.m.?

Write your answer in 24-hour clock form.

☐☐ : ☐☐

9) HAV bank holds £4 million in assets.

GFR bank holds three times as many assets as HAV bank.

FVE bank holds £5,238,238 in assets.

What is the total amount of assets held by the 3 banks?

£ ☐☐ , ☐☐☐ , ☐☐☐

10) The 4 internal angles in a quadrilateral are 117°, 129°, $X°$ and $2X°$.

What is the value of $X°$?

☐☐ °

11) 17 students took a test. The mean score was 73%, the highest score was 85%, the median score was 71% and the range was 36%.

What was the lowest score?

☐☐ %

12) What is the difference in area between a rectangle with a length of 7 cm and a width of 8 cm and a triangle with a base of 4 cm and a height of 28 cm?

☐ cm²

Test 21

INSTRUCTIONS

You have 6 minutes to complete the following section.

You have 12 questions to complete within the time given.

Write the correct answer in the boxes provided (one digit per box).

EXAMPLE

One quarter of the students in Class A have brown eyes. If there are 52 students in Class A, how many do not have brown eyes?

The answer is:

| 3 | 9 |

Number of students who have brown eyes = 52 ÷ 4 = 13
Number of students who do not have brown eyes = 52 − 13 = 39

---

(1) Calculate ((4 × 18) − (17 + 55)) × 12

(2) How many tenths are there in $4\frac{2}{5}$?

(3) Write 0.95 as a fraction in its lowest terms.

(4) Calculate 44 mm × 8

⬜.⬜⬜⬜ m

(5) What is the range of these values?

0.032, 0.033, 0.031, 0.321, 0.08

⬜.⬜⬜

(6) How many cubes with a side length of 2 cm can fit inside a cube with a side length of 4 cm?

(7) A square and a rectangle have the same area. The rectangle has a width of 4 cm and a length of 9 cm.

What is the perimeter of the square?

cm

(8) 4 out of every 5 days in June are sunny.

How many more sunny days than non-sunny days are there in June?

HELPFUL TIP: *Think about how many days there are in June in total.*

(9) Henry divided his stamp collection into 5 equal piles and sold 10% of the stamps in each pile. Each stamp was sold for £4.50 and he received a total of £90.

How many stamps did Henry have left in his collection?

(10) A cuboid measures 6 cm by 8 cm by 16 cm.

How many cubes with a volume of 0.125 cm³ could fit into the cuboid?

☐,☐☐☐

(11) Every day, 10 litres of water are removed from a tank and 7.5 litres are added to the tank.

If the tanks currently holds 65 litres of water, after how many weeks will it hold 30 litres of water?

weeks

(12) What is the size of the acute angle between the hour and the minute hand on a clock at 4:30 p.m.?

°

Test 22

 You have 6 minutes to complete the following section.
You have 12 questions to complete within the time given.

Write the correct answer in the boxes provided.

EXAMPLE

Calculate (7 × 9) + (6 × 8)

The answer is:

| 1 | 1 | 1 |

(7 × 9) + (6 × 8) = 63 + 48 = 111

---

(1) Calculate 44,000 − 73,222 + 50,383

☐☐,☐☐☐

(2) $4 ÷ 18 = 24 ÷ P$

What is the value of P?

☐☐☐

(3) How many times greater is $4\frac{1}{2}$ than 1.5?

☐

(4) What number is 7 less than half of 89?

☐☐.☐

(5) Georgina bought 8 lemons for £1.34.

What is the cost of 20 lemons?

£ ☐.☐☐

Page 46

(6) Which shape has exactly 2 pairs of parallel sides and exactly 2 lines of symmetry?

(7) Hannah is 143 cm tall and Rebecca is 1.3 m tall.

The total height of Hannah, Rebecca and Emma is 4,500 mm.

How tall is Emma?

[ ][ ][ ] cm

(8) A nail weighs 2.5 g and a screw weighs 4 g.

What is the total weight of 160 nails and 200 screws?

[ ].[ ] kg

(9) 2 cubes, each with a side length of 6 cm, are placed together to form a cuboid.

What is the surface area of the cuboid?

[ ][ ][ ] cm²

HELPFUL TIP: *Remember that a cube has 6 identical square faces.*

(10) The cost of an online course rose from £125 to £137.50.

What was the percentage increase in the cost of the course?

[ ][ ] %

(11) A regular hexagon with a perimeter of 79.5 cm is enlarged by a scale factor of 6.

What is the length of 1 side of the enlarged hexagon?

[ ][ ].[ ] cm

(12) $X + Y = 43$

$X - Y = 1$

What is the value of Y?

[ ][ ]

Test 23

 You have 6 minutes to complete the following section.
You have 12 questions to complete within the time given.

Write the correct answer in the boxes provided.

EXAMPLE

1 orange costs £0.65. What is the cost of 6 oranges?

The answer is:

| 3 | 9 | 0 | p |

£0.65 = 65 p
65 p × 6 = 390 p

---

(1) Calculate 364 ÷ 13

(2) Calculate $\frac{7}{10} - \frac{1}{5}$

Write your answer as a fraction in its lowest terms.

(3) What is the square root of $\frac{1}{5}$ of 125?

(4) Calculate £8.42 × $\frac{3}{2}$

£

(5) Nina thinks of a number, squares it, adds 7 and then squares it again.

The answer is 121.

What number did Nina think of?

6 4 apples weigh 32 g, 46 g, 21 g and X g respectively.

The mean weight of the 4 apples is 33 g.

What is the value of X?

7 15% of the students in a class have brown eyes. Half of the remaining students have blue eyes.

What percentage of students in the class do not have blue eyes? ☐☐.☐ %

8 Calculate $(\frac{1}{6} \times \frac{7}{8}) \div (\frac{4}{5} \times \frac{3}{5})$

Write your answer as a fraction in its lowest terms.

9 Vic has 54 sweets. He gives 9 sweets to Bill so that they now have the same number of sweets.

How many sweets did Bill originally have?

10 If 17th August is a Monday, what day is 10th September in the same year?

11 $\frac{1}{6}$ of the fish in a tank have green scales and $\frac{1}{3}$ of the fish in the tank have blue scales.

What decimal fraction of the fish in the tank have neither green nor blue scales? ☐.☐

12 What is 175% of $\frac{1}{2}$ of 0.9 of 180? ☐☐☐.☐☐

Test 24

INSTRUCTIONS

 You have 6 minutes to complete the following section.
You have 12 questions to complete within the time given.

Write the correct answer in the boxes provided (one digit per box).

EXAMPLE

One quarter of the students in Class A have brown eyes. If there are
52 students in Class A, how many do not have brown eyes?

The answer is:

3	9

Number of students who have brown eyes = 52 ÷ 4 = 13
Number of students who do not have brown eyes = 52 − 13 = 39

---

① Calculate $(\frac{3}{4} \times 640) - (\frac{2}{3} \times 90) + (\frac{1}{6} \times 60)$

② Calculate 223×18

③ Calculate $0.5 \div 0.2$

④ Calculate $-19 + 13 - 8 + 45$

⑤ The angles in a quadrilateral are $E°$, $\frac{E°}{2}$, $E°$ and $\frac{E°}{2}$.
What is the value of E?

(6) How many times greater is the product of 9 and 11 than the product of 16.5 and 2?

☐

(7) $\frac{3}{8}$ of a litre of cola costs 45 p.

What is the cost of $1\frac{1}{2}$ litres of cola?

£ ☐ . ☐ ☐

HELPFUL TIP: *It may be useful to first calculate the cost of 1 litre of cola.*

(8) 7 blue balls, 1 green ball and 3 red balls are placed in a bag.

1 ball is selected at random.

What is the probability that the selected ball is not red?

$\dfrac{☐}{☐}$

(9) 8 chairs cost £170.66.

What is the cost of 1 chair, rounded to the nearest whole pence?

£ ☐ ☐ . ☐ ☐

(10) There were 187 pictures for sale in a gallery.

After 7 weeks, there were 138 pictures remaining for sale.

How many pictures were sold on average per week?

☐

(11) A gym offers a monthly membership of £19.99 and an annual membership of £220.

Over a 2-year period, how much cheaper is the annual membership than the monthly membership?

£ ☐ ☐ . ☐ ☐

(12) Carl's mean score in 6 tests is 57.

After 7 tests, his mean score is 52.

What was Carl's score in the 7th test?

☐ ☐

Test 25

 You have 6 minutes to complete the following section.
You have 12 questions to complete within the time given.

Write the correct answer in the boxes provided (one digit per box).

EXAMPLE

Calculate $(7 \times 9) + (6 \times 8)$

The answer is:

| 1 | 1 | 1 |

$(7 \times 9) + (6 \times 8) = 63 + 48 = 111$

---

(1) Calculate $0.08 \div 2$

◻.◻◻

(2) How many fifths are there in 13.6?

◻◻

(3) Calculate $17.883 \times 2 \times 0.1$

◻.◻◻◻

(4) Round 349.4092 to 2 decimal places.

◻◻◻.◻◻

(5) Calculate $\frac{1}{6} \times (\frac{1}{6} \times (\frac{1}{6} \times 216))$

◻

(6) What is the mode of the following values?

0.3, 0.2, 0.5, 0.3, 0.2, 0.1, 0.9, 0.2, 0.3, 0.6, 0.2,
0.7, 0.1, 0.3, 0.5, 0.2

☐.☐

(7) X is 350% greater than Y.

If $X = 1,050$, what is the value of Y?

☐☐☐

(8) The average height of 11 plants is 11.5 cm.

What is the total height of the 11 plants?

☐☐☐.☐ cm

(9) Kelly has 3 dresses, 2 hats and 5 pairs of shoes.

If she chooses to wear 1 dress with 1 hat and 1 pair of shoes, how many different combinations are possible?

☐☐

(10) Triangle A is right-angled. It has a perimeter of 24 cm and the lengths of its 2 longest sides are 10 cm and 8 cm.

2 identical versions of Triangle A are placed together to form a rectangle.

What is the area of the rectangle?

☐☐ cm²

(11) The cost of a holiday in pounds is given by the expression $110R + 360E$.

R represents the number of travellers and E represents the length of the holiday in weeks.

What is the cost of a holiday for 4 travellers for 7 days?

£☐☐☐

(12) There are 7 black balls and 8 yellow balls in a bag.
1 black ball and 2 yellow balls are removed from the bag.

A ball is then selected at random.

What is the probability that the selected ball is yellow?

Write your answer as a fraction in its lowest terms.

$\frac{☐}{☐}$

Test 26

INSTRUCTIONS

 You have 6 minutes to complete the following section.
You have 12 questions to complete within the time given.

Write the correct answer in the boxes provided (one digit per box).

EXAMPLE

1 orange costs £0.65. What is the cost of 6 oranges?

The answer is:

| 3 | 9 | 0 | p |

£0.65 = 65 p
65 p × 6 = 390 p

---

(1) Calculate 45 km ÷ 2

☐,☐☐☐,☐☐☐ cm

(2) Calculate $\frac{1}{18} + \frac{4}{9}$

Write your answer as a fraction in its lowest terms.

$\frac{\square}{\square}$

(3) $T + 13 = 2T + 1$

What is the value of T?

☐☐

(4) 12% of $B = \frac{1}{2}$ of 72

What is the value of B?

☐☐☐

(5) Each week a snake's length is 1.2 times its length in the previous week.

If a snake is 1.4 m long, how long will it be in 3 weeks' time?

Round your answer to the nearest whole cm.

◻◻◻ cm

(6) Calculate $\frac{3}{8} + \frac{1}{2} + \frac{54}{64}$

Write your answer as a top-heavy fraction in its lowest terms.

$\frac{\boxed{}\boxed{}}{\boxed{}\boxed{}}$

(7) What is $\frac{1}{7}$ rounded to 2 decimal places?

◻.◻◻

(8) How many fewer days were there in the first 3 months of 2017 than the last 3 months of 2016?

◻

(9) The price of a theatre ticket decreased from £25 to £21.50.

What was the percentage decrease in price?

◻◻ %

(10) $x^2 + x^2 + x^2 = 75$

What is the value of x?

◻

(11) What is the size of the reflex angle between the hour and the minute hand on a clock at 17:45?

◻◻◻.◻ °

(12) The coordinates of 3 vertices of a square are (4, 8), (4, 12) and (8, 12).

What are the coordinates of the midpoint of the square?

(◻ , ◻◻)

Test 27

 You have 6 minutes to complete the following section.
You have 12 questions to complete within the time given.

Write the correct answer in the boxes provided (one digit per box).

EXAMPLE

One quarter of the students in Class A have brown eyes. If there are
52 students in Class A, how many do not have brown eyes?

The answer is:

| 3 | 9 |

Number of students who have brown eyes = 52 ÷ 4 = 13
Number of students who do not have brown eyes = 52 − 13 = 39

---

(1) Calculate £8 × 1.2

£ ☐ . ☐ ☐

(2) Calculate 0.02 × 0.01

☐ . ☐ ☐ ☐

(3) Calculate 12 days ÷ 4

☐ , ☐ ☐ ☐ minutes

(4) Calculate $\frac{3}{5}$ of a litre

☐ ☐ ☐ ml

(5) Add the probability of rolling a 2 on a fair dice to the
probability of tossing a coin and it landing on heads.

Write your answer as a fraction in its lowest terms.

☐
—
☐

6 Rearrange these digits to make the largest possible even number.

8, 2, 9, 5, 1

⬚⬚,⬚⬚⬚

7 Tins of paint cost £4.50 each. A current deal offers every 5th tin free if 4 are bought at full price.

Using this deal, what is the total cost of 10 tins of paint?

£⬚⬚

8 A sausage contains 3 g of fat for every 1.2 g of protein.

If there are 12 g of protein in the sausage, what is the difference in weight of the protein in the sausage and the fat in the sausage?

⬚⬚ g

9 For any values b and c, $b \star c = bc + b - c$

If $b = 7$ and $c = 4$, what is the value of $b \star c$?

⬚⬚

10 $\frac{2}{5}$ of the cost of a bag is £14.

What number must £14 be multiplied by to give the cost of 10 bags?

⬚⬚

11 How many more lines of symmetry does a regular pentagon have than a square?

⬚

12 The average weight of 7 snails is 56 g.

3 other snails weigh 54 g, 60 g and 40 g respectively.

What is the average weight of all 10 snails?

⬚⬚.⬚ g

Test 28

 INSTRUCTIONS

 You have 6 minutes to complete the following section.
You have 12 questions to complete within the time given.

Write the correct answer in the boxes provided.

EXAMPLE

Calculate $(7 \times 9) + (6 \times 8)$

The answer is:

| 1 | 1 | 1 |

$(7 \times 9) + (6 \times 8) = 63 + 48 = 111$

---

(1) How much greater is 4,392,883 than 3,928,281?

☐☐☐,☐☐☐

(2) $\frac{7}{8}$ of $X = 2P$

If $P = 35$, what is the value of X?

☐☐

(3) Calculate 453×2.1

☐☐☐.☐

(4) Calculate $5(7^2) - 44$

☐☐☐

(5) It takes Joe 17 minutes to travel to work in the morning and $\frac{2}{3}$ of an hour to travel back home in the afternoon.

Assuming Joe works 5 days per week, how long does he spend travelling to and from work per week?

☐☐☐ minutes

(6) A 20 p coin weighs 1.6 g.

What is the value of a pile of 20 p coins with a total
weight of 48 g?

£ ☐

(7) Reena turns 135° clockwise and then 225° anticlockwise

She now faces south.

In which direction was she originally facing?

☐

(8) $70 < F < 80$

When F is divided by 7, the remainder is 5.

What is the value of F?

☐☐

(9) The average of 5 consecutive numbers is 18.

What is the sum of the largest and the smallest number?

☐☐

(10) Gina is 7 years older than Sam. Katy is 2 years younger
than Sam. Last year, Katy was 8.

How old will Gina be next year?

☐☐

(11) $x^7 - 12 = -11$

What is the value of x?

☐

(12) Lewis thinks of a number, subtracts 6, multiplies by 8 and
then divides by 5.

The answer is 3.2.

What number did Lewis think of?

☐

HELPFUL TIP: *Start with the answer and work backwards, inversing the calculations.*

Test 29

 You have 6 minutes to complete the following section.
You have 12 questions to complete within the time given.

Write the correct answer in the boxes provided.

EXAMPLE

1 orange costs £0.65. What is the cost of 6 oranges?

The answer is:

| 3 | 9 | 0 | p |

£0.65 = 65 p
65 p × 6 = 390 p

---

(1) Calculate $\frac{1}{8} + \frac{5}{8} - \frac{3}{8}$

Write your answer as a decimal.

☐ . ☐ ☐ ☐

(2) How many seconds are there in $\frac{1}{5}$ of an hour?

☐ ☐ ☐ seconds

(3) Calculate 78 ml $- \frac{1}{20}$ of a litre

☐ . ☐ ☐ ☐ l

(4) Calculate 17% of 50

☐ . ☐

(5) 1 euro has the same value as 0.84 pounds.

What is the cost in pounds of a shirt that costs 12 euros ?

£ ☐ ☐ . ☐ ☐

6 Express 2 mm to 6 metres as a ratio in its lowest terms.

☐ : ☐ , ☐ ☐ ☐

7 A bag contains 2 stones that each weigh 23 g and 3 stones that each weigh 45 g.

What is the average weight per stone in the bag?

Round your answer to the nearest whole gram.

☐ ☐ g

8 Don has 3 times as many cookies as Elvis. Hannah has half as many cookies as Ruben. Elvis has 3 times as many cookies as Ruben.

If Hannah has 22 cookies, how many cookies do they have altogether?

☐ ☐ ☐

9 Plant A is 56 cm in height, rounded to the nearest cm.

Plant B is exactly 3.5 cm taller than Plant A.

What is the maximum possible height of Plant B, rounded to 2 decimal places?

☐ ☐ . ☐ ☐

10 A faulty watch runs 2 minutes slow per hour.

If the watch shows the correct time at 14:30, what time will it show 3.5 hours later?

☐ ☐ : ☐ ☐

11 Colin faces east and turns 45° clockwise. He then turns to face the opposite direction.

In which direction does he now face?

☐

12 7 buses transport some tourists to a museum. Each bus can hold a maximum of 28 tourists. On average, there are 7 empty seats in each bus.

How many tourists are there in total?

☐ ☐ ☐

Test 30

 You have 6 minutes to complete the following section.
You have 12 questions to complete within the time given.

Write the correct answer in the boxes provided (one digit per box).

EXAMPLE

One quarter of the students in Class A have brown eyes. If there are
52 students in Class A, how many do not have brown eyes?

The answer is:

3	9

Number of students who have brown eyes = 52 ÷ 4 = 13
Number of students who do not have brown eyes = 52 − 13 = 39

---

(1) How many times greater is 7^2 than 7?

(2) Calculate $\frac{3}{4} \div \frac{1}{6}$

Write your answer as a fraction in its lowest terms.

(3) What is $5\frac{7}{8}$ rounded to 2 decimal places?

(4) Calculate 8,316 p ÷ 18

£ ☐.☐☐

(5) A bus begins its journey with 46 girls and 23 boys on board.
At the first stop, half the girls and 12 boys get off and 17 girls
and 11 boys get on.

How many more girls than boys are there now on the bus?

(6) A pizza is divided into 9 equal slices.

The area of the pizza is 45.36 cm².

What is the combined area of 4 of the slices of pizza?

☐☐.☐☐ cm²

(7) The total length of 4 sides of a hexagon is 43.2 cm.

The remaining sides of the hexagon each measure 12.2 cm.

What is the mean length per side of the hexagon?

Round your answer to 2 decimal places.

☐☐.☐☐ cm

(8) Which number comes next in this sequence?

134, 117, 101, 86, ?

☐☐

(9) What is the product of the number of faces of a cuboid and the number of edges of a hexagonal prism?

☐☐☐

(10) A gazelle eats $\frac{1}{2}$ kg of grass per day.

How much grass does the gazelle eat in total in October and November?

☐☐.☐ kg

(11) 240 children were asked what their favourite colour was and the results were displayed using a pie chart.

The segment on the pie chart representing the number of children whose favourite colour was blue measured 18°.

How many children chose blue as their favourite colour?

☐☐

(12) A cardboard rectangle has a length of 7 cm and width of 6 cm.

A square with an area of 16 cm² is cut out of 1 corner of the rectangle.

What is the perimeter of the remaining cardboard?

☐☐ cm

Notes

Answers

Key abbreviations: °C: degrees centigrade, cm: centimetre, d.p.: decimal place, g: gram, h: hour, kg: kilogram, km: kilometre, kph: kilometres per hour, l: litre, m: metre, min: minutes, ml: millilitre, mm: millimetre

Test 1

Q1 **2.25**

Q2 **5.81**

Q3 $\frac{1}{12}$

Q4 **60**
1.5 kg = 1,500 g
1,500 g ÷ 25 g = 60

Q5 **70,409**

Q6 **4**
5 × 5 = 25
6 × 6 = 36
7 × 7 = 49
8 × 8 = 64

Q7 **290**
4 hours = 4 × 60 minutes = 240 minutes
$\frac{5}{6}$ of an hour = $\frac{5}{6}$ × 60 minutes = 50 minutes
240 minutes + 50 minutes = 290 minutes

Q8 **84**
Cost of 1 apple = £1.47 ÷ 7 = £0.21
Cost of 4 apples = £0.21 × 4 = £0.84
= 84 pence

Q9 **348**
There are 12 months in 1 year.
Annual cost = 12 × £29 = £348

Q10 **38.40**
20% of £32 = £6.40
New price = £32 + £6.40 = £38.40

Q11 **110**
Width = 44 cm ÷ 4 = 11 cm = 110 mm

Q12 **40**
360° ÷ 9 = 40°

Test 2

Q1 **1,904**

Q2 **10**
√100 = 10

Q3 $\frac{5}{12}$
$\frac{2}{3} - \frac{1}{4} = \frac{8}{12} - \frac{3}{12} = \frac{5}{12}$

Q4 **4,000**
44 m ÷ 11 = 4 m = 400 cm
= 4,000 mm

Q5 **9**
Factors of 36: 1, 36, 2, 18, 3, 12, 4, 9, 6

Q6 **81**
Sequence consists of square numbers in ascending order.

Q7 **8**
2 cm × 2 cm = 4 cm²
So side length of square is 2 cm
Perimeter = 2 cm × 4 = 8 cm

Q8 **24**
1% of £25 = £0.25; 4% = £0.25 × 4 = £1
96% = 100% − 4% = £25 − £1 = £24

Q9 **1.5**
Radius of Circle B = 50 cm ÷ 4 = 12.5 cm
Diameter of Circle B = 12.5 cm × 2 = 25 cm
Radius of Circle A = 25 cm × 3 = 75 cm
Diameter of Circle A = 75 cm × 2 = 150 cm
150 cm = 1.5 m

Q10 **28.8**
Distance covered in 10 seconds
= 400 m ÷ 5 = 80 m
Distance covered in 1 minute
= 80 m × 6 = 480 m
Distance covered in 1 hour = 480 m × 60
= 28,800 m = 28.8 km

Q11 **28**
00:25 to 02:45 is 140 minutes
140 minutes ÷ 20 minutes = 7
Number of chicks born = 4 × 7 = 28

Q12 **(−4, 9)**

Test 3

Q1 **28,044**

Q2 **15**

Q3 **66**
(3 × 7) + (5 × 9) = 21 + 45 = 66

Q4 **34.28**

Q5 **110**
10 + 100 = 110

Q6 **14.1**
Perimeter = 2.35 cm × 6 = 14.1 cm

Q7 **89**
Size of 3rd angle = $180° - 67° - 24° = 89°$

Q8 **120**
8 + 7 = 15; so there are 15 parts in total
1 part = 56 ÷ 7 = 8 sweets
15 parts = 8 sweets × 15 = 120 sweets

Q9 **2**

Q10 **2**
Factors of 36: 1, 36, 2, 18, 3, 12, 4, 9, 6
Factors of 48: 1, 48, 2, 24, 3, 16, 4, 12, 6, 8
Factors of 54: 1, 54, 2, 27, 3, 18, 6, 9
Factors of both 36 and 48 but not 54: 4, 12

Q11 **7.75**
Cost of a child ticket = (£2.17 ÷ 7) × 2
= £0.31 × 2 = £0.62
Total cost = £0.62 + £0.62 + £2.17 +
£2.17 + £2.17 = £7.75

Q12 **15**
Total number of children = 5 × 4 = 20
Number of female children = $\frac{3}{4}$ × 20 = 15

Test 4

Q1 $\frac{71}{3}$

Q2 **8,221**

Q3 **8**
71 − 63 = 8

Q4 **20**
$\frac{5}{1} \div \frac{1}{4} = \frac{5}{1} \times \frac{4}{1} = \frac{20}{1} = 20$

Q5 **40,842**
10% of 45,380 = 4,538
45,380 − 4,538 = 40,842

Q6 **17**
March has 31 days so there are 15 days
of the festival in March and 2 days in April.
15 + 2 = 17

Q7 **6**
Height = (area ÷ base) × 2 =
(36 ÷ 12) × 2 = 6 cm

Q8 **2.31**
330 ml × 7 = 2,310 ml = 2.31 litres

Q9 **70**
Medicine taken per day = 10 ml + 15 ml
= 25 ml
Capsules consumed per day
= 25 ÷ 5 = 5
Capsules consumed per fortnight
= 5 × 14 = 70

Q10 **12.24**
Cheapest possible combination is
11 packs and 1 single teabag.
Total cost = (11 × £1.10) + £0.14
= £12.10 + £0.14 = £12.24

Q11 **200**
11% of savings = £22
1% of savings = £22 ÷ 11 = £2
Total savings = £2 × 100 = £200

Q12 **22**
Tuesday temperature = $29° - 10° = 19°$
Thursday temperature = $19° + 3° = 22°$

Test 5

Q1 **320,000**

Q2 **13**

Q3 $\frac{7}{20}$
$\frac{28}{80} = \frac{7}{20}$

Q4 **5.6**

Q5 **36.6**
Total capacity = 12.2 litres × 3 = 36.6 litres

Q6 **567**
$9^2 = 81$
81 × 7 = 567

Q7 **89.28**
Radius = 17.856 cm ÷ 2 = 8.928 cm
= 89.28 mm

Q8 **382.5**
Distance travelled in 1 hour =
45 km × 2 = 90 km
Distance travelled in $4\frac{1}{4}$ hours =
4.25 × 90 km = 382.5 km

Q9 **48**
Number of green toads = 128 ÷ 2 = 64
Number of green toads without yellow
spots = $\frac{3}{4}$ × 64 = 48

Q10 **30**
Actual distance = 3 cm × 1,000,000
= 3,000,000 cm = 30,000 m = 30 km

Q11 **12**

Prime factors of 96: $2 \times 2 \times 2 \times 2 \times 2 \times 3$

Prime factors of 84: $2 \times 2 \times 3 \times 7$

Prime factors of 60: $2 \times 2 \times 3 \times 5$

Highest common factor = $2 \times 2 \times 3 = 12$

Q12 **23.10**

Cost for 1 day = £3 + (30 min × £0.01) = £3.30

Total cost for 1 week = £3.30 × 7 = £23.10

Test 6

Q1 **114.1**

$T = 68.9 + 45.2 = 114.1$

Q2 $\frac{5}{8}$

$62.5\% = \frac{62.5}{100} = \frac{5}{8}$

Q3 **10.01**

Q4 **352,517**

Q5 **120**

Sale price is 80% of original price.

80% = £96

10% = £96 ÷ 8 = £12

Original price = £12 × 10 = £120

Q6 **27**

3rd angle = $180° - 90° - 27° = 63°$

So $27°$ is the smallest of the 3 angles.

Q7 **36**

Number of tiles to fit along floor length = 300 ÷ 50 = 6

Number to fill whole area = 6 × 6 = 36

Q8 **600**

Distance travelled in 1 minute

= 2.5 km × 4 = 10 km

Distance travelled in 1 hour

= 10 km × 60 = 600 km

Q9 **19**

One set of two-pence and five-pence coins has a value of 7 pence.

Number of sets = 133 pence ÷ 7 pence = 19

If there are 19 sets, there must be 19 five-pence coins.

Q10 **346.50**

10% of £350 = £35

Tuesday price = £350 − £35 = £315

10% of £315 = £31.50; Wednesday price

= £315 + £31.50 = £346.50

Q11 **54**

Side length of square = $\sqrt{81}$ cm^2 = 9 cm

Length of each rectangle = 9 cm

Width of each rectangle = 4.5 cm

Perimeter of 1 rectangle = 9 cm + 9 cm + 4.5 cm + 4.5 cm = 27 cm

Total perimeter = 27 cm × 2 = 54 cm

Q12 **57**

Test 7

Q1 **97**

81 + 16 = 97

Q2 **77,376**

Q3 **14,000,006**

Q4 **560**

56 kg = 56,000 g; 56,000 g ÷ 100 = 560 g

Q5 $\frac{1}{2}$

Number of even outcomes = 3

(2, 4 and 6 are even)

Total number of outcomes = 6

Probability = $\frac{3}{6} = \frac{1}{2}$

Q6 **4.56**

2,000 g = £15.20

1,000 g = £15.20 ÷ 2 = £7.60

100 g = £7.60 ÷ 10 = £0.76

600 g = £0.76 × 6 = £4.56

Q7 **68.4**

Perimeter = 22.8 cm × 3 = 68.4 cm

Q8 **3.185**

Total weight = 455 kg × 7 = 3,185 kg

= 3.185 tonnes

Q9 **63**

Number of parts = 3 + 4 + 2 + 5 = 14

Two-pence and ten-pence coins have an even value.

So $\frac{9}{14}$ of the coins have an even value.

$\frac{9}{14} \times 98 = 63$

Q10 **20**

Total cost = £1.21 × 4 = £4.84

Note value = cost + change

= £4.84 + £15.16 = £20

Q11 **56**

$\frac{1}{3}$ of 24 hours = 8 hours

8 hours × 7 = 56 hours

Q12 **90**

15 spiders can spin 9 webs in 6 minutes.
15 spiders can spin 90 webs in 60 minutes.

Test 8

Q1 **9.76**

976 cm = 9.76 m

Q2 **7**

$W \times 110 = 10 \times 77$
$110W = 770$
$W = 770 \div 110 = 7$

Q3 **93**

£0.93 = 93 p

Q4 **1,430,000**

Q5 **southeast**

$135° = 90° + 45°$
North to east is 90°
East to southeast is 45°

Q6 **4:04**

$1\frac{1}{3}$ hours = 1 hour and 20 minutes
1 hour 20 minutes before 5:24 p.m. is
4:04 p.m.

Q7 **64**

$4^3 = 4 \times 4 \times 4 = 64$

Q8 **24.5**

Width = 7 cm ÷ 2 = 3.5 cm
Area = 7 cm × 3.5 cm = 24.5 cm^2

Q9 **10.125**

Side length of the square = 18 cm ÷ 4
= 4.5 cm
Area of the square = 4.5 cm × 4.5 cm
= 20.25 cm^2
Half the area of the circle = 20.25 cm^2 ÷ 2
= 10.125 cm^2

Q10 **kite**

Q11 **10**

10 cm × 10 cm × 10 cm = 1,000 cm^3

Q12 **200**

75% of 1,000 = 750
0.8 of 750 = 600

$\frac{1}{2}$ of 600 = 300

$\frac{2}{3}$ of 300 = 200

Test 9

Q1 **6**

973 seconds − 613 seconds = 360
seconds
360 seconds ÷ 60 = 6 minutes

Q2 **433,664**

Q3 **14**

$\sqrt{49} = 7$; $2 \times 7 = 14$

Q4 **200**

30% of C = 60
10% of C = 60 ÷ 3 = 20
$C = 20 \times 10 = 200$

Q5 $\frac{1}{2}$

Number of prime outcomes = 3
(2, 3 and 5 are prime numbers)
Total number of outcomes = 6
Probability = $\frac{3}{6} = \frac{1}{2}$

Q6 **Saturday**

16th October to 7th November is 22 days
22 days = 3 weeks and 1 day
So 7th November must be 1 day after
Friday → Saturday

Q7 **8**

Number of litres drained every hour =
7 × 4 = 28 litres
224 litres ÷ 28 = 8 hours

Q8 **(2, 13)**

The coordinates must be the only other
possible combination of x and y.

Q9 **56**

Surface area = (2 × 10 cm × 3 cm) +
(2 × 10 cm × 2 cm) + (2 × 2 cm × 3 cm)
= 60 cm^2 + 40 cm^2 + 12 cm^2 = 112 cm^2
Surface area not painted red = 112 cm^2 ÷ 2
= 56 cm^2

Q10 **2**

Q11 **550**

1 mm is ten times smaller than 1 cm
Actual distance = 5,500 m ÷ 10 = 550 m

Q12 $\frac{6}{13}$

Total number of cards = 15 + 12 − 1
= 26
Total number of red cards = 12
Probability = $\frac{12}{26} = \frac{6}{13}$

Test 10

Q1 **156.96**

Q2 **6,626**
7,000 m − 374 m = 6,626 m

Q3 **45**
$5\frac{5}{8} = \frac{45}{8}$

Q4 **36**
7 weeks = 7 × 7 days = 49 days
49 days − 13 days = 36 days

Q5 **Any number greater than 3.45 but less than 3.46.**
Any of the following: 3.451, 3.452, 3.453, 3.454, 3.455, 3.456, 3.457, 3.458, 3.459

Q6 **6**
Numbers in Group F: 20, 22, 24, 26, 28, 30

Q7 **144**
$\frac{5}{9}$ of the animals = 80 animals
$\frac{1}{9}$ = 80 animals ÷ 5 = 16 animals
Total number of animals = 16 × 9 = 144

Q8 **10.4**
Area = 13 m × 0.8 m = 10.4 m²

Q9 **19**
11 people in front of him and 7 people behind him.
Total number of people = 11 + 7 + 1 = 19

Q10 **43**
Otto: 43 seconds
Toby: 39 seconds
Alan: 47 seconds
Average time = (43 + 39 + 47) ÷ 3
= 129 ÷ 3 = 43 seconds

Q11 **27**
$6* = (4 × 6) + \frac{6}{2} = 24 + 3 = 27$

Q12 **0.051**
0.051 − 0.05 = 0.001
0.05 − 0.005 = 0.045
0.055 − 0.05 = 0.005
0.493 − 0.05 = 0.443
0.05 − 0.048 = 0.002
0.05 − 0.045 = 0.005

Test 11

Q1 **120**
$9Q$ = 433 + 647 = 1,080
Q = 1,080 ÷ 9 = 120

Q2 **47**

Q3 **4.898**

Q4 **128**
121 + (16 − 9) = 121 + 7 = 128

Q5 **1**
2 is the only even prime number.

Q6 **130**
Length of model = 1,040 cm ÷ 8 = 130 cm

Q7 **51.127**
34.29 × 2 = 68.58
68.58 − 17.453 = 51.127

Q8 **147**
Sum of other angles = 360° − 93° − 120°
= 147°

Q9 **789,499**

Q10 **12**
Perimeter of hexagon = 8 cm × 6 = 48 cm
Side length of octagon = 48 cm ÷ 8 = 6 cm
Length of 2 sides = 6 cm + 6 cm = 12 cm

Q11 **700**
15% = 105
1% = 105 ÷ 15 = 7
100% = 7 × 100 = 700

Q12 **1,171.50**
Cost of 2 seats = £35.50
Cost of 6 seats = £35.50 × 3 = £106.50
Total cost = £106.50 × 11 = £1,171.50

Test 12

Q1 **2,600**
2.6 m = 260 cm = 2,600 mm

Q2 **1.956**
1.4 l + 0.345 l + 0.211 l = 1.956 l

Q3 **10,989**

Q4 **1.1**
0.5 + 0.6 = 1.1

Q5 **4**
−18 °C + 22 °C = 4 °C

Q6 **5**
Number of months in year with 31 days = 7
Number of leap years from 1921 to 1969
= 12; 12 − 7 = 5

Q7 **210**

Length of 1 piece = 350 cm ÷ 50 = 7 cm

Length of 3 pieces = 7 cm × 3 = 21 cm

= 210 mm

Q8 **18**

Work backwards: 48 ÷ 3 = 16

16 − 7 = 9; 9 × 2 = 18

Q9 **8**

Value in pounds = 9.2 euros ÷ 1.15 = £8

Q10 **64**

Difference = 1,357 − 1,293 = 64

Q11 **22**

This year, Fred is 6; last year, Fred was 5.
Last year, Belle was 5 × 4 = 20; next year
Belle will be 22.

Q12 **7**

Test 13

Q1 **7,500**

Q2 **210**

Q3 **0**

$T × 12 = 33 × 0$

$12T = 0; T = 0$

Q4 **105,285**

Q5 **90**

3 lions eat 6 kg of meat in 30 minutes

18 kg ÷ 6 kg = 3 kg

Time taken = 30 minutes × 3 = 90 minutes

Q6 **−45,431**

Q7 **2**

A cube has 6 faces; $\frac{2}{3}$ of 6 = 4

Faces not painted blue = 6 − 4 = 2

Q8 **500**

Volume = 25 m × 10 m × 2 m = 500 m³

Q9 **0**

Since D is a positive number, the
right-hand side of the equation must
equal 0. Therefore, C must also equal 0.

Q10 **4**

1,000 km ÷ 25 km = 40

Distance on map = 40 × 1 mm = 40 mm

= 4 cm

Q11 **400**

Length of each smaller rectangle = 4 cm

Width of each smaller rectangle = 1 cm

Perimeter of each smaller rectangle =
4 cm + 4 cm + 1 cm + 1 cm = 10 cm

Total perimeter = 4 × 10 cm = 40 cm

= 400 mm

Q12 **0.407**

Metres above sea level = 757 m − 350 m

= 407 m = 0.407 km

Test 14

Q1 **9.96**

Q2 **4**

$\frac{1}{8}$ of 64 = 8; 8 ÷ 2 = 4

Q3 **260**

$\frac{R}{2}$ = 130

$R = 130 × 2 = 260$

Q4 **35,856**

Q5 **0.24**

Weight of 2 apples = 220 g × 2 = 440 g

Weight of 3 lemons = 800 g − 440 g

= 360 g

Weight of 1 lemon = 360 g ÷ 3 = 120 g

Weight of 2 lemons = 120 g × 2 = 240 g

= 0.24 kg

Q6 **456**

The smallest value coin (1 pence) must be
used as many times as possible.

456 × 1 p = 456 p = £4.56

Q7 **7:30**

360° = 1 full turn

810° ÷ 360° = 2.25 = $2\frac{1}{4}$ turns

$2\frac{1}{4}$ turns = 2 hours 15 minutes

2 hours 15 minutes after 5:15 p.m.
is 7:30 p.m.

Q8 **12**

16 ÷ 4 = 4; 4 × 7 = 28

Number of blue buttons = 16

Number of red buttons = 28

Difference = 28 − 16 = 12

Q9 **northeast**

90° anticlockwise turn from north to
face west

225° anticlockwise turn from west to
face northeast

Q10 $\frac{3}{10}$

Fraction of board painted blue =
1 − 0.5 − 0.2 = 0.3
Probability = 0.3 = $\frac{3}{10}$

Q11 **0.075**

2,000 ml ÷ 175 ml = 11 remainder 75
75 ml = 0.075 l

Q12 **2,340**

$\frac{3}{5}$ of an hour = $\frac{3}{5}$ × 60 minutes
= 36 minutes
$\frac{5}{8}$ of 2 hours = $\frac{5}{8}$ × 120 minutes
= 75 minutes
Difference = 75 minutes − 36 minutes
= 39 minutes
39 minutes = (39 × 60) seconds
= 2,340 seconds

Test 15

Q1 **30**

15% of 200 cm = 30 cm

Q2 **56**

1 fortnight = 14 days; 4 × 14 = 56 days

Q3 **7 30**

45 hours ÷ 6 = 7.5 hours
7.5 hours = 7 h 30 min

Q4 $\frac{19}{24}$

$\frac{4}{6} + \frac{1}{8} = \frac{16}{24} + \frac{3}{24} = \frac{19}{24}$

Q5 **2**

19, 23

Q6 **721**

Change received = £10 − £0.93 − £0.93
− £0.93 = £7.21 = 721 p

Q7 $\frac{1}{4}$

Finn eats $\frac{1}{2}$ of $\frac{1}{2}$ of the cake, so he eats $\frac{1}{4}$
Fraction remaining = $1 - \frac{1}{2} - \frac{1}{4} = \frac{1}{4}$

Q8 **670**

536 cm ÷ 8 = 67 cm = 670 mm

Q9 **44**

Ascending order: 12, 23, 32, 43, 45, 65,
76, 87
Median = (43 + 45) ÷ 2 = 88 ÷ 2 = 44

Q10 **18**

£10 − £5.14 = £4.86
Number of pens = 486 p ÷ 27 p = 18

Q11 **74**

Distance covered by Train A = 2 × 43 =
86 km
Distance covered by Train B = 2.5 × 64 =
160 km
Difference = 160 km − 86 km = 74 km

Q12 **1,350**

Height of Triangle A = (225 ÷ 15) × 2 =
15 × 2 = 30 cm
Height of Triangle B = 3 × 30 cm = 90 cm
Base of Triangle B = 2 × 15 cm = 30 cm
Area of Triangle B = $\frac{1}{2}$ × 30 cm × 90 cm
= 1,350 cm²

Test 16

Q1 **6,174.60**

Q2 **59.2**

592 mm = 59.2 cm

Q3 **12.566**

14 km − (0.234 km + 1.2 km) =
14 km − 1.434 km = 12.566 km

Q4 **36.2**

Q5 **0.078632**

Each term is ten times smaller than the
previous term so the decimal point moves
one position to the left each time.

Q6 **54.6**

Side length of hexagon = 27.3 cm ÷ 3
= 9.1 cm
Perimeter of hexagon = 9.1 cm × 6
= 54.6 cm

Q7 **1,219.3**

1,220 cm − 0.7 cm = 1,219.3 cm

Q8 **1.21**

Price after 1 year = 1.1 × £1 = £1.10
Price after 2 years = 1.1 × £1.10 = £1.21

Q9 **0.09**

Number of notes = £870 ÷ £10 = 87
Total thickness = 87 × 0.01 mm =
0.87 mm = 0.087 cm = 0.09 cm (2 d.p.)

Q10 **1,000**

Scale is 1 cm:10,000 cm
Length of playing field = 0.5 cm × 10,000
= 5,000 cm = 50 m
Width of playing field = 0.2 cm × 10,000
= 2,000 cm = 20 m
Area = 50 m × 20 m = 1,000 m²

Q11 **(3, −8)**

Q12 **90**

3 builders take 30 minutes so 1 builder will take three times as long.
30 minutes × 3 = 90 minutes

Test 17

Q1 **360**

$B = 2 \times 45 = 90$
$4B = 4 \times 90 = 360$

Q2 **20**

$25 - 5 = 20$

Q3 **15,204**

Q4 **$6\frac{3}{7}$**

$45 \div 7 = 6$ remainder 3

Q5 **2,000**

Number of turns = 4,000 m ÷ 2 m = 2,000

Q6 **58**

7 less than double 18 = 36 − 7 = 29
29 × 2 = 58

Q7 **112**

14:23 to 14:51 is 28 minutes
28 minutes ÷ 7 minutes = 4
So the number of flies doubles 4 times
7 × 2 × 2 × 2 × 2 = 112

Q8 **134**

1 hundredth = 0.01
134 × 0.01 = 1.34

Q9 **0.96**

5 Toms = 18 × 20 p = 360 p
1 Tom = 360 p ÷ 5 = 72 p
4 Toms = 72 p × 4 = 288 p
1 Tim = 288 p ÷ 9 = 32 p
3 Tims = 32 p × 3 = 96 p = £0.96

Q10 **7.50**

18 ÷ 6 = 3; 4 × 3 = 12
Number of eggs needed = 12
Number of boxes needed = 2
Cost = 2 × £3.75 = £7.50

Q11 **24**

Sequence: +11, −6, +11, −6…
−35 + 11 = −24

Q12 **14**

$1\frac{6}{8} = \frac{14}{8}$
$\frac{14}{8} \div \frac{1}{8} = 14$

Test 18

Q1 **45,700,000**

Q2 **18**

$4\frac{1}{2} = 4\frac{2}{4} = \frac{18}{4}$

Q3 **540**

£0.90 × 6 = £5.40 = 540 pence

Q4 **104**

64 + 44 − 4 = 104

Q5 **0**

A square number cannot be prime.

Q6 **117**

£2.25 × 52 = £117

Q7 **0**

$180° - 93° = 87°$
$87° \div 2 = 43.5°$
$43.5° - 43.5° = 0$

Q8 **25**

Total number of parts = 1 + 3 + 1 = 5
Value of each part = £125 ÷ 5 = £25
Grace receives 3 parts = £25 × 3 = £75
Gina and Gail receive 2 parts = £25 × 2 = £50
£75 − £50 = £25

Q9 **46.75**

Area to cover = 2 × (9 m × 6 m) = 108 m²
Number of tins needed = 108 m² ÷ 10 m² = 10.8
So 11 tins needed, rounded up.
Cost = 11 × £4.25 = £46.75

Q10 **(−4, −14)**

Coordinates of Point C = (−4, 14)
Coordinates of Point E = (−4, −14)

Q11 **189**

13 out of every 17 apples are non-rotten.
Number of non-rotten apples = (84 ÷ 4) × 13 = 21 × 13 = 273
Difference = 273 − 84 = 189 apples

Q12 **360**

Test 19

Q1 **10**

1, 80, 2, 40, 4, 20, 5, 16, 8, 10

Q2 **137**

$7X = 1,000 - 41 = 959$
$X = 959 \div 7 = 137$

Q3   21,982
23,000 − 1,018 = 21,982

Q4   18
There are 37 halves in $18\frac{1}{2}$
There are 19 quarters in $4\frac{3}{4}$
37 − 19 = 18

Q5   53.25
$\frac{1}{2}$ of 6 = 3; $\frac{2}{3}$ of 21 = 14; sum = 3 + 14 = 17
$\frac{1}{4}$ of 1 = 0.25; $\frac{7}{8}$ of 80 = 70; sum = 0.25 +
70 = 70.25
70.25 − 17 = 53.25

Q6   150.75
33.5 km × 4.5 = 150.75 km

Q7   72
Length = 12 cm; width = 6 cm
Area = 12 cm × 6 cm = 72 cm²

Q8   1
532,400 p ÷ 1,000,000 = 0.5324 p
0.5324 p rounded to the nearest pence
is 1 p.

Q9   490
Total number of wheels =
5 × ((7 × 8) + (7 × 4) + (7 × 2))
= 5 × (56 + 28 + 14) = 5 × 98 = 490

Q10   384
15:12 to 16:01 is 49 minutes
49 minutes ÷ 7 minutes = 7
So the quantity of cells doubles 7 times.
$3 \times 2^7 = 3 \times 128 = 384$

Q11   150
Total cost in 11 instalments = £15 × 11
= £165
110% of total 6-instalment cost = £165
10% of total 6-instalment cost =
£165 ÷ 11 = £15
100% of total 6-instalment cost =
£15 × 10 = £150

Q12   48,000
Total parts = 52 + 48 = 100
Difference in votes cast as a fraction =
$\frac{52}{100} - \frac{48}{100} = \frac{4}{100} = \frac{1}{25}$
$\frac{1}{25} \times 1,200,000 = 48,000$

Test 20

Q1   76.33

Q2   12.293
12,543 ml − 250 ml = 12,293 ml = 12.293 l

Q3   6,503.9
6,504 − 0.1 = 6,503.9

Q4   7
10, 12, 14, 16, 18, 20, 22

Q5   3.80
£2 + £1 + £0.50 + £0.20 + £0.10 = £3.80

Q6   37
Mean = (45 + 65 + 19 + 19) ÷ 4 =
148 ÷ 4 = 37

Q7   3
Cost of 1 bag = £0.40 × 3 = £1.20
Cost of $2\frac{1}{2}$ bags = 2.5 × £1.20 = £3

Q8   21:11
$2\frac{4}{5}$ hours is 2 hours and 48 minutes
2 hours 48 minutes before 11:59 p.m. is
9:11 p.m.
9:11 p.m. is equivalent to 21:11 in
24-hour clock format.

Q9   21,238,238
Assets held by GFR = £4,000,000 × 3 =
£12,000,000
Total assets = £12,000,000 + £4,000,000 +
£5,238,238 = £21,238,238

Q10   38
117° + 129° + X° + $2X$° = 360°
$3X$° = 360° − 117° − 129° = 114°
X° = 114° ÷ 3 = 38°

Q11   49
Lowest score = highest score − range =
85% − 36% = 49%

Q12   0
Area of rectangle = 7 cm × 8 cm = 56 cm²
Area of triangle = $\frac{1}{2}$ × 4 cm × 28 cm =
2 × 28 = 56 cm²
Difference = 56 cm² − 56 cm² = 0 cm²

Test 21

Q1   0
(72 − 72) × 12 = 0 × 12 = 0

Q2   44
$4\frac{2}{5} = 4\frac{4}{10} = \frac{44}{10}$

Q3 $\frac{19}{20}$
$0.95 = \frac{95}{100} = \frac{19}{20}$

Q4 **0.352**

44 mm × 8 = 352 mm = 35.2 cm = 0.352 m

Q5 **0.29**

Range = 0.321 − 0.031 = 0.29

Q6 **8**

Volume of cube with 2 cm side =
2 cm × 2 cm × 2 cm = 8 cm³
Volume of cube with 4 cm side =
4 cm × 4 cm × 4 cm = 64 cm³
Number of cubes = 64 ÷ 8 = 8

Q7 **24**

Area of rectangle = 4 cm × 9 cm = 36 cm²
Side length of square = √36 cm² = 6 cm
Perimeter of square = 6 cm × 4 = 24 cm

Q8 **18**

Number of days in June = 30
Number of sunny days = $\frac{4}{5}$ × 30 = 24
Number of non-sunny days = 30 − 24 = 6
Difference = 24 − 6 = 18

Q9 **180**

Total number of stamps sold =
£90 ÷ £4.50 = 20
10% of collection = 20 stamps
90% of collection = 20 stamps × 9 =
180 stamps

Q10 **6,144**

Volume of cuboid = 6 cm × 8 cm × 16 cm
= 768 cm³
0.125 cm³ = $\frac{1}{8}$ cm³
Number of cubes = 768 ÷ $\frac{1}{8}$ = 768 × 8
= 6,144

Q11 **2**

Daily reduction = 10 litres − 7.5 litres =
2.5 litres
65 litres − 30 litres = 35 litres
Number of days = 35 litres ÷ 2.5 litres = 14
14 days = 2 weeks

Q12 **45**

Size of angle between 2 adjacent numbers
on a clock = 360° ÷ 12 = 30°
Acute angle between 2 hands at 4:30 p.m.
is 1.5 times this distance.
30° × 1.5 = 45°

Test 22

Q1 **21,161**

Q2 **108**

$\frac{4}{18} = \frac{24}{P}$
$4P = 18 × 24$
So $P = 18 × 6 = 108$

Q3 **3**

$4\frac{1}{2} = 4.5$
4.5 ÷ 1.5 = 3

Q4 **37.5**

89 ÷ 2 = 44.5
44.5 − 7 = 37.5

Q5 **3.35**

20 ÷ 8 = 2.5
Cost of 20 lemons = 2.5 × £1.34 = £3.35

Q6 **Rectangle**

Q7 **177**

Emma's height = 450 cm − 143 cm −
130 cm = 177 cm

Q8 **1.2**

Total weight = (160 × 2.5 g) + (200 × 4 g)
= 400 g + 800 g = 1,200 g = 1.2 kg

Q9 **360**

Surface area of 1 face of cube =
6 cm × 6 cm = 36 cm²
Surface area of cuboid = Surface area of
1 face of cube × 10
10 × 36 cm² = 360 cm²

Q10 **10**

Difference in price = £137.50 − £125 =
£12.50
Percentage increase = (£12.50 ÷ £125) ×
100 = 0.1 × 100 = 10%

Q11 **79.5**

Perimeter of enlarged hexagon =
79.5 cm × 6 = 477 cm
Length of 1 side = 477 cm ÷ 6 =
79.5 cm

Q12 **21**

X = 22; Y = 21

Test 23

Q1 **28**

Q2 $\frac{1}{2}$

$\frac{7}{10} - \frac{1}{5} = \frac{7}{10} - \frac{2}{10} = \frac{5}{10} = \frac{1}{2}$

Q3 **5**

$\frac{1}{5}$ of 125 = 25; $\sqrt{25}$ = 5

Q4 **12.63**

£8.42 ÷ 2 = £4.21
£4.21 × 3 = £12.63

Q5 **2**

Work backwards: $\sqrt{121}$ = 11; 11 − 7 = 4
$\sqrt{4}$ = 2

Q6 **33**

Total weight of the apples =
4 × 33 g = 132 g
X = 132 − 32 − 46 − 21 = 33

Q7 **57.5**

Students with blue eyes =
(100% − 15%) ÷ 2 = 42.5%
Students without blue eyes =
100% − 42.5% = 57.5%

Q8 $\frac{175}{576}$

$\frac{1}{6} \times \frac{7}{8} = \frac{7}{48}$
$\frac{4}{5} \times \frac{3}{5} = \frac{12}{25}$
$\frac{7}{48} \div \frac{12}{25} = \frac{7}{48} \times \frac{25}{12} = \frac{175}{576}$

Q9 **36**

Number of sweets Vic and Bill each have
now = 54 − 9 = 45
Number of sweets Bill originally had =
45 − 9 = 36

Q10 **Thursday**

17th August to 10th September is 24 days.
24 days is 3 weeks and 3 days.
So 10th September must be 3 days after
Monday → Thursday

Q11 **0.5**

Neither blue nor green scales = $1 - \frac{1}{6} - \frac{1}{3}$
$= 1 - \frac{1}{6} - \frac{2}{6} = \frac{3}{6} = \frac{1}{2} = 0.5$

Q12 **141.75**

0.9 of 180 = 162
$\frac{1}{2}$ of 162 = 81
175% of 81 = 141.75

Test 24

Q1 **430**

480 − 60 + 10 = 430

Q2 **4,014**

Q3 **2.5**

Q4 **31**

Q5 **120**

$E° + E° + \frac{E°}{2} + \frac{E°}{2} = 3E°$
$3E° = 360°$
$E° = 360° ÷ 3 = 120°$

Q6 **3**

9 × 11 = 99; 16.5 × 2 = 33
99 ÷ 33 = 3

Q7 **1.80**

$\frac{1}{8}$ of a litre costs 45 p ÷ 3 = 15 p
1 litre costs 15 p × 8 = £1.20
$1\frac{1}{2}$ litres costs £1.20 × 1.5 = £1.80

Q8 $\frac{8}{11}$

Total number of balls = 7 + 1 + 3 = 11
Total number of non-red balls = 11 − 3 = 8
Probability = $\frac{8}{11}$

Q9 **21.33**

Cost of 1 chair = £170.66 ÷ 8 = £21.3325
= £21.33 (rounded to the nearest whole
pence)

Q10 **7**

Number of pictures sold = 187 − 138 = 49
Average sold per week = 49 ÷ 7 = 7

Q11 **39.76**

Cost over 2 years with monthly
membership = 24 × £19.99 = £479.76
Cost over 2 years with annual
membership = 2 × £220 = £440
Difference in price = £479.76 − £440 =
£39.76

Q12 **22**

Total marks scored after 6 tests =
57 × 6 = 342
Total marks scored after 7 tests =
52 × 7 = 364
Score in 7th test = 364 − 342 = 22

Test 25

Q1 **0.04**

Q2 **68**

$13.6 = 13\frac{3}{5} = \frac{68}{5}$

Q3 **3.5766**

Q4 **349.41**

Q5 **1**

216 ÷ 6 = 36; 36 ÷ 6 = 6; 6 ÷ 6 = 1

Q6 **0.2**
0.2 occurs the most frequently.

Q7 **300**
X is 3.5 times greater than Y.
$Y = 1{,}050 \div 3.5 = 300$

Q8 **126.5**
Total height = 11 × 11.5 cm = 126.5 cm

Q9 **30**
Number of combinations = 3 × 2 × 5 = 30

Q10 **48**
Length of shortest side of rectangle =
24 cm − 10 cm − 8 cm = 6 cm
Area of rectangle = 6 cm × 8 cm = 48 cm²

Q11 **800**
7 days = 1 week
$R = 4$; $E = 1$
Cost of holiday = (110 × 4) + (360 × 1) =
440 + 360 = £800

Q12 $\frac{1}{2}$
Total number of balls = 7 + 8 − 1 − 2 = 12
Total number of yellow balls = 8 − 2 = 6
Probability = $\frac{6}{12} = \frac{1}{2}$

Test 26

Q1 **2,250,000**
45 km ÷ 2 = 22.5 km = 22,500 m
= 2,250,000 cm

Q2 $\frac{1}{2}$
$\frac{1}{18} + \frac{4}{9} = \frac{1}{18} + \frac{8}{18} = \frac{9}{18} = \frac{1}{2}$

Q3 **12**
$T + 13 = 2T + 1$
$2T - T = 13 - 1$
$T = 12$

Q4 **300**
$\frac{1}{2}$ of 72 = 36
12% of B = 36
1% of B = 36 ÷ 12 = 3
$B = 100 × 3 = 300$

Q5 **242**
Length = 1.2 × 1.2 × 1.2 × 1.4 m =
2.4192 m = 241.92 cm
241.92 cm rounded to the nearest whole
cm is 242 cm.

Q6 $\frac{55}{32}$
$\frac{24}{64} + \frac{32}{64} + \frac{54}{64} = \frac{110}{64} = \frac{55}{32}$

Q7 **0.14**
$1 \div 7 = 0.14285\ldots = 0.14$ (2 d.p.)

Q8 **2**
First 3 months of 2017 = 31 + 28 + 31 =
90 days
Last 3 months of 2016 = 31 + 30 + 31 =
92 days
92 − 90 = 2 days

Q9 **14**
Decrease = £25 − £21.50 = £3.50
Percentage decrease = $\frac{£3.50}{£25} × 100 =$
0.14 × 100 = 14%

Q10 **5**
$3x^2 = 75$
$x^2 = 75 \div 3 = 25$
$x = \sqrt{25} = 5$

Q11 **262.5**
Size of angle between 2 adjacent numbers
on clock face = 360° ÷ 12 = 30°
Reflex angle between hands at 17:45 is
8.75 times this amount.
8.75 × 30° = 262.5°

Q12 **(6, 10)**
Side length of square is 4 units.
Midpoint coordinates = ((4 + 2), (8 + 2))
= (6, 10)

Test 27

Q1 **9.60**

Q2 **0.0002**

Q3 **4,320**
3 days = 24 hours × 3 = 72 hours
72 hours = 72 × 60 minutes =
4,320 minutes

Q4 **600**
$\frac{3}{5}$ of a litre = $\frac{3}{5}$ of 1,000 ml = 600 ml

Q5 $\frac{2}{3}$
Probability of rolling a 2 = $\frac{1}{6}$
Probability of landing on heads = $\frac{1}{2}$
$\frac{1}{6} + \frac{1}{2} = \frac{1}{6} + \frac{3}{6} = \frac{4}{6} = \frac{2}{3}$

Q6 **98,512**

Q7 **36**
Total cost of 5 tins = 4 × £4.50 = £18
Total cost of 10 tins = £18 × 2 = £36

Q8 **18**

$12 \text{ g} \div 1.2 \text{ g} = 10$

Amount of fat in sausage = $3 \text{ g} \times 10 = 30 \text{ g}$

$30 \text{ g} - 12 \text{ g} = 18 \text{ g}$

Q9 **31**

$7 \bigstar 4 = (7 \times 4) + 7 - 4 = 28 + 7 - 4 = 31$

Q10 **25**

Number = $10 \div \frac{2}{5} = 10 \times \frac{5}{2} = \frac{50}{2} = 25$

Q11 **1**

Lines of symmetry of regular pentagon = 5

Lines of symmetry of square = 4

Difference = $5 - 4 = 1$

Q12 **54.6**

Total weight = $(7 \times 56 \text{ g}) + 54 \text{ g} + 60 \text{ g} + 40 \text{ g} = 546 \text{ g}$

Average weight = $546 \text{ g} \div 10 = 54.6 \text{ g}$

Test 28

Q1 **464,602**

$4,392,883 - 3,928,281 = 464,602$

Q2 **80**

$\frac{7}{8}X = 2 \times 35 = 70$

$\frac{1}{8}X = 70 \div 7 = 10$

$X = 10 \times 8 = 80$

Q3 **951.3**

Q4 **201**

$5(7^2) - 44 = (5 \times 49) - 44 = 245 - 44 = 201$

Q5 **285**

$17 \text{ minutes} \times 5 = 85 \text{ minutes}$

$\frac{1}{3}$ of an hour = 20 minutes so $\frac{2}{3}$ of an hour = 40 minutes

$40 \text{ minutes} \times 5 = 200 \text{ minutes}$

$85 \text{ minutes} + 200 \text{ minutes} = 285 \text{ minutes}$

Q6 **6**

$48 \text{ g} \div 1.6 \text{ g} = 30$

Value of coins in pile = $20 \text{ p} \times 30 = 600 \text{ p} = £6$

Q7 **west**

$225° - 135° = 90°$

So in total she turns $90°$ anticlockwise. A $90°$ anticlockwise turn from west will leave her facing south.

Q8 **75**

$75 \div 7 = 10$ remainder 5

Q9 **36**

Numbers are 16, 17, 18, 19 and 20

Sum of the largest and smallest numbers = $16 + 20 = 36$

Q10 **19**

This year, Katy is 9; this year, Sam is 11. This year Gina is 18, next year Gina will be 19.

Q11 **1**

$x^7 = -11 + 12 = 1$

$x = 1$

Q12 **8**

Work backwards: $3.2 \times 5 = 16$

$16 \div 8 = 2; 2 + 6 = 8$

Test 29

Q1 **0.375**

$\frac{1}{8} + \frac{5}{8} - \frac{3}{8} = \frac{6}{8} - \frac{3}{8} = \frac{3}{8} = 0.375$

Q2 **720**

$\frac{1}{5}$ of an hour = $\frac{1}{5} \times 60$ minutes = 12 minutes

12 minutes = 12×60 seconds = 720 seconds

Q3 **0.028**

$\frac{1}{20}$ of a litre = $1,000 \text{ ml} \div 20 = 50 \text{ ml}$

$78 \text{ ml} - 50 \text{ ml} = 28 \text{ ml} = 0.028 \text{ l}$

Q4 **8.5**

17% of 50 = $\frac{17}{100} \times 50 = \frac{17}{2} = 8.5$

Q5 **10.08**

Cost in pounds = 12 euros $\times 0.84 = £10.08$

Q6 **1:3,000**

$6 \text{ m} = 600 \text{ cm} = 6,000 \text{ mm}$

$2:6,000 = 1:3,000$

Q7 **36**

Total weight = $23 \text{ g} + 23 \text{ g} + 45 \text{ g} + 45 \text{ g} + 45 \text{ g} = 181 \text{ g}$

Average weight = $181 \text{ g} \div 5 = 36.2 \text{ g}$

36.2 g rounded to the nearest whole gram is 36 g

Q8 **594**

Hannah: 22 cookies

Ruben: $22 \times 2 = 44$ cookies

Elvis: $44 \times 3 = 132$ cookies

Don: $132 \times 3 = 396$ cookies

Total = $22 + 44 + 132 + 396 = 594$ cookies

Q9 59.99

Maximum height of Plant A is 56.49 cm (2 d.p.)

Maximum height of Plant B is 56.49 cm + 3.5 cm = 59.99 cm

Q10 17:53

Minutes slow in 3.5 hours =
3.5 × 2 minutes = 7 minutes

Actual time 3.5 hours after 14:30 is 18:00

7 minutes before 18:00 is 17:53

Q11 northwest

45° clockwise from east is southeast.

Opposite direction (180°) to southeast is northwest.

Q12 147

Average number of tourists per bus =
28 − 7 = 21

Total number of tourists = 21 × 7 = 147

Test 30

Q1 7

$7^2 = 49$; $49 \div 7 = 7$

Q2 $\frac{9}{2}$

$\frac{3}{4} \div \frac{1}{6} = \frac{3}{4} \times \frac{6}{1} = \frac{18}{4} = \frac{9}{2}$

Q3 5.88

$5\frac{7}{8} = 5.875 = 5.88$ (2 d.p.)

Q4 4.62

8,316 pence ÷ 18 = 462 pence = £4.62

Q5 18

Number of girls = (46 ÷ 2) + 17 =
23 + 17 = 40

Number of boys = 23 − 12 + 11 = 22

Difference = 40 − 22 = 18

Q6 20.16

Area of 1 slice = 45.36 cm² ÷ 9 = 5.04 cm²

Area of 4 slices = 5.04 cm² × 4 = 20.16 cm²

Q7 11.27

Perimeter of hexagon = 43.2 cm +
12.2 cm + 12.2 cm = 67.6 cm

Mean length per side = 67.6 cm ÷ 6 =
11.266… cm = 11.27 cm (2 d.p.)

Q8 72

Sequence: −17, −16, −15, −14…

86 −14 = 72

Q9 108

Number of faces of a cuboid = 6

Number of edges of a hexagonal prism = 18

18 × 6 = 108

Q10 30.5

Number of days in October and November = 31 + 30 = 61

Amount of grass eaten = 61 × 0.5 kg = 30.5 kg

Q11 12

360° ÷ 18° = 20

Number of children who chose blue = 240 ÷ 20 = 12

Q12 26

Perimeter of rectangle = 7 cm + 7 cm + 6 cm + 6 cm = 26 cm

Cutting a square with side length of 4 cm out of 1 corner will not change the perimeter.

Perimeter of remaining cardboard = 26 cm

Notes

Score Sheet

Below is a score sheet to track your results over multiple attempts. One mark is available for each question in the tests.

Test	Pages	Date of first attempt	Score out of 12	Date of second attempt	Score out of 12	Date of third attempt	Score out of 12
Test 1	4–5						
Test 2	6–7						
Test 3	8–9						
Test 4	10–11						
Test 5	12–13						
Test 6	14–15						
Test 7	16–17						
Test 8	18–19						
Test 9	20–21						
Test 10	22–23						
Test 11	24–25						
Test 12	26–27						
Test 13	28–29						
Test 14	30–31						
Test 15	32–33						
Test 16	34–35						
Test 17	36–37						
Test 18	38–39						
Test 19	40–41						
Test 20	42–43						
Test 21	44–45						
Test 22	46–47						
Test 23	48–49						
Test 24	50–51						
Test 25	52–53						
Test 26	54–55						
Test 27	56–57						
Test 28	58–59						
Test 29	60–61						
Test 30	62–63						